COUNTER-INTELLIGENCE

Protecting Us from the Realities of Roswell

RICHARD KESTNER

Counterintelligence: Protecting Us from the Realities of Roswell

Published by K235 Media
High Rolls Mountain Park, NM

DISCLAIMER
Names, characters, businesses, places, events, locales, and incidents are either the products of the author's imagination or used in a fictitious manner. Any resemblance to actual persons, living or dead, or actual events is purely coincidental.

ISBN: 978-0-578-28611-2

HISTORY / Military / Intelligence & Espionage

Cover and interior design by CreativeReflections.Design, copyright owned by Richard Kestner.

ISBN 13-978-0-578-28611-2

DEDICATION

This book is dedicated to anyone who has
wondered about what really happened near
Roswell, New Mexico, in July 1947. Searching for
the truth is always admirable, sometimes necessary,
and seemingly never easy.

ACKNOWLEDGMENTS

TO ACKNOWLEDGE everyone who contributed to this book is a challenge, especially since it has been in the creative process for almost fifteen years.

First and foremost, thank you to my high school sweetheart, life partner, and adventuring companion Durema. Thanks for your patience, support, assistance, and gentle persuasive method. We could not have completed this effort together had you not been actively involved from the very start to the end.

We gratefully acknowledge the invaluable assistance of Mr. Stanton Friedman, especially with specific answers to specific questions all along the way. We both still have electronic records on our phones from communications with him while we were at the National Archives at College Park, Maryland. We remember fondly his openness and friendliness at the annual Roswell event. We miss him and the trail he blazed for all of us to follow, even though we might not agree with everything he wrote.

We gratefully acknowledge the United States Air Force. We are deeply indebted to Airmen, Airwomen, and civilian personnel of the service. Special thanks to the Air Force History Research Agency at Maxwell Air Force Base in Montgomery, Alabama; the United

States Air Force Academy in Colorado Springs, Colorado; and retired Air Force individuals associated with the Walker Air Force Base Museum in Roswell, New Mexico.

We would be remiss if we did not mention the publishing industry folks who treated us with courtesy, information, and encouragement. We contacted more than twenty agents, companies, and individuals. Almost all went above and beyond the simple efforts to win our business and make us informed patrons. Special thanks to the My Word Publishing team: Stephanie, Patrice, and Melissa. Special thanks also to the 'volunteer' proofing team.

LOS ALAMOS
X
SANDIA LABS
X
CONFIDENTIAL
TRINITY SITE
X
WHITE
SANDS
PROVING
GROUNDS
ROSWE
NEW MEX

CONTENTS

LOS ALAMOS
SANDIA LABS
CONFIDENTIAL
TRINITY SITE
WHITE SANDS PROVING GROUNDS
NEW MEX

CHAPTER 1
Inspiration and Motivation

It is a Saturday afternoon in Roswell, New Mexico, and I am sitting among the overflow crowd in attendance for the opening ceremony of the new Walker Air Force Base Museum. The Walker Aviation Museum Foundation has put together a modest two-room display at the Roswell International Air Center terminal that is the city's airport these days. They have also erected a monument at the entrance to the terminal "... dedicated to the men and women who valiantly served this nation while stationed at Walker Air Force Base."

There are a lot of old timers in attendance who have shown up to swap stories and relive some of their glory days at what was, at one time, the premier Strategic Air Command base of the United States Air

Force or Army Air Forces as they were known prior to September 1947. I am here, because I am curious to see how they will handle the Incident of 1947.

The program is led by Brigadier General Douglas Murray, US Air Force Retired who is currently Academic Dean at the New Mexico Military Institute (on the other end of Roswell from the former Walker Air Force Base) in addition to being one of the key members of the Museum Foundation. The New Mexico Military Institute is well represented on the dais by its color guard, regimental band, and various military school faculty members. Major General Chris Adams, US Air Force Retired is the keynote speaker, and Roswell Mayor Del Jurney is here representing the city.

The crowd is animated and active while being polite and orderly. In addition to the museum opening, a monument is being dedicated. It is no accident that the event has been scheduled for September 18, the sixty-third birthday of the United States Air Force. I feel like a spy among the retired military folks, the conservative local community members, and the well-heeled general public in attendance.

I rise for the national anthem and applaud politely when appropriate. I avoid most conversation with another attendee on my left, having found a seat at the end of a row. The speeches are pretty routine for this kind of event, but I am listening carefully for any reference whatsoever to the Incident.

The *Enola Gay* and *Bockscar* (sometimes called *Bock's Car)*, historic aircraft that dropped the atomic bombs on Japan, are mentioned as well as the fact that the first atomic weapons of the Cold War were most likely here in 1947. Outlying Atlas missile silos commanded from Walker Air Force Base are mentioned. The World War II B-29 Bomber and the B-50 of the Cold War are mentioned. The huge B-36 and the still-flying B-52 are all mentioned. Some of the units that once called the base home are mentioned including the 509th Bomb Group (formerly the 509th Composite Group during WWII), which dropped the atomic bombs on Japan. But there is no mention whatsoever of the Incident.

It is as I expected, but as I sit and reflect, letting the crowd exit ahead of me, I am suddenly struck by the thought that the United States Air Force, through its representatives and retired officers, is continuing to deny the historic Incident of 1947. It is a US Air Force rewrite – a rewrite of omission. If we don't address it, it doesn't exist. If we rewrite the history with the omission, it never occurred. A rewrite of omission means there is no history of the July 1947 Roswell Incident.

On my drive home, my brain is racing. I am already anticipating adding my own personal narrative along with the many folks who have addressed the Incident from their view. As I ponder my personal history and the history of the US Air Force Roswell Incident, I mentally organize the strategy that the US Air Force uses to continue their handling of the Incident. Four words come to mind: *deception, distraction, denigration,* and *omission.* From July 1947 to the present day, the US Air Force counterintelligence campaign strategy to omit the Incident continues.

Roswell. Like unused herbs and spices in a kitchen cabinet, the name is slowly losing its potency over time. Ten or twenty years ago, you could mention Roswell, and the reaction was universal. People all over the world were familiar with the story of the 1947 capture of a "flying disk" by the United States Army Air Forces at Roswell Army Air Field in New Mexico.

The decline of the name recognition is not a natural progression or a fluke. It is the successful result of a seventy-five-year-long, ongoing US Air Force counterintelligence (CI) campaign to protect the information and knowledge associated with that July 1947 event, which came to be called the Roswell Incident.

After almost eight years in Roswell, multiple visits to the National Archives and Records Administration

II facility in College Park, Maryland, and a visit to the US Air Force History Research Agency facility at Maxwell Air Force Base in Montgomery, Alabama, this is my deductive reasoning conclusion based on the evidence.

There are hundreds, maybe even thousands of books in many different languages about the Roswell Incident. When one mentions New Mexico in foreign lands, the state is not generally known. Mentioning Albuquerque might get a response, because of the role the city played in the successful television series *Breaking Bad*. But Santa Fe, Las Cruces, or even Alamogordo seldom receive recognition. Roswell. That name has been instantly acknowledged almost anywhere in the world, but that recognition is fading.

Like many folks, I struggled to understand how the US Air Force could issue a press release saying they had captured a flying saucer only to almost immediately report they made a mistake – it was a weather balloon.

When I moved to Roswell in 2007, I determined that I would investigate and decide for myself where the truth lay. This book relates my personal findings, critical thoughts, observations, and conclusions from 2007 to 2022. I will use US Air Force, Air Force, and USAF here, fully cognizant that in July

1947, then Colonel William H. "Butch" Blanchard and the folks of the 509th Bomb Group (successor to the WWII 509th Composite Group) were part of the United States Army Air Forces stationed at Roswell Army Air Field. It wasn't until September 1947, almost three months after the Incident, that the US Air Force was birthed as a separate service of the new Department of Defense. In 1949, Roswell Army Air Field was renamed Walker Air Force Base.

If you have read this far and anticipate reading about aliens and crashed spacecraft, you will probably be disappointed. This book is about the facts of the Incident itself and the readily discernable counterintelligence (CI) campaign the US Air Force continues to conduct to protect the information concerning the Roswell Incident. This book lays out the facts and evidence I have uncovered. Based on research and reasoning, you will learn whether the original "flying disk" claim by the Roswell Army Air Field is the most likely story that rings true – or if it is not.

I don't know beyond any doubt what happened in July 1947 near Roswell, New Mexico. I don't know if it was extraterrestrials. I don't know if it was the data-gathering, research balloon called Project Mogul. But I am pretty sure, and the Air Force has admitted, it was not a weather balloon. In the future, I fear that the rewrite of history to omit the Incident will forever prevent us from knowing what really happened. And now, as the seventy-fifth anniversary of the Incident approaches, it appears that the rewrite of history – by omission – may have already happened.

This book is about my personal journey to resolve this mystery for myself as well as doing some small part to prevent the rewrite of history by omission that might forever close us off from learning what really happened. This is not an academic endeavor with footnotes, cited sources, and the like, although there is a resource list at the end. This is not a scientific or engineering-based effort with lots of measurements, timelines, hardware descriptions, and the like. This is not a "he said/she said" reporting of hearsay and third-party statements with an attempt to qualify the statements.

This is a personal statement of observations about how the US government – in particular the US Army Air Forces (later the US Air Force) – has continued its attempt to rewrite the history from that original "flying saucer" press release as part of a seventy-five-year, ongoing counter-intelligence campaign.

FAST FORWARD TO NOVEMBER 23, 2021

The Department of Defense issued a press release on this autumn Tuesday. It is establishing a new organization to be known as AOIMSG under the USD (I&S). Other parts of the release include AOIMEXEC, SUA, DNI, and UAP. Deputy Secretary of Defense Kathleen Hicks in coordination with the Director of National Intelligence (DNI) directed the Under Secretary of Defense for Intelligence and Security (USDI&S) to establish the Airborne Object Identification and Management Synchronization Group (AOIMSG) as successor to the US Navy's Unidentified Aerial Phenomena Task Force (UAPTF). The AOIMSG's mission is stated as: "The AOIMSG will synchronize efforts across the Department and the broader US government to detect, identify, and attribute objects of interest in Special Use Airspace (SUA), and to assess and mitigate any associated threats to safety of flight and national security." Additionally, the AOIMSG will have oversight from the USDI&S with the help of the Airborne Object Identification and Management Executive Council (AOIMEXEC), presumably to be established by the USDI&S.

Can you follow all this? What does it mean? This development in the ongoing Department of Defense response to UFOs since 1947 can be viewed as either a promise of more transparency or a higher level of continuation of the US Air Force's counterintelligence campaign to preserve and protect the information concerning the Roswell Incident.

Changing UFO to UAP (Unidentified Aerial Phenomena), broadening the number of services involved from one to two (US Air Force and US Navy), assigning women to leadership roles, and committing DoD command-level involvement might portend preparations for finally allowing a complete disclosure of the 1947 Roswell Incident.

On the other hand, this might just be another delaying tactic to keep us from the truth. In recent years, the USAF issued a *Case Closed* report related to Roswell that was intended to close the Roswell case, once and for all. Is this November 2021 press release just the latest counterintelligence effort? Keep reading – let's follow the trail of evidence.

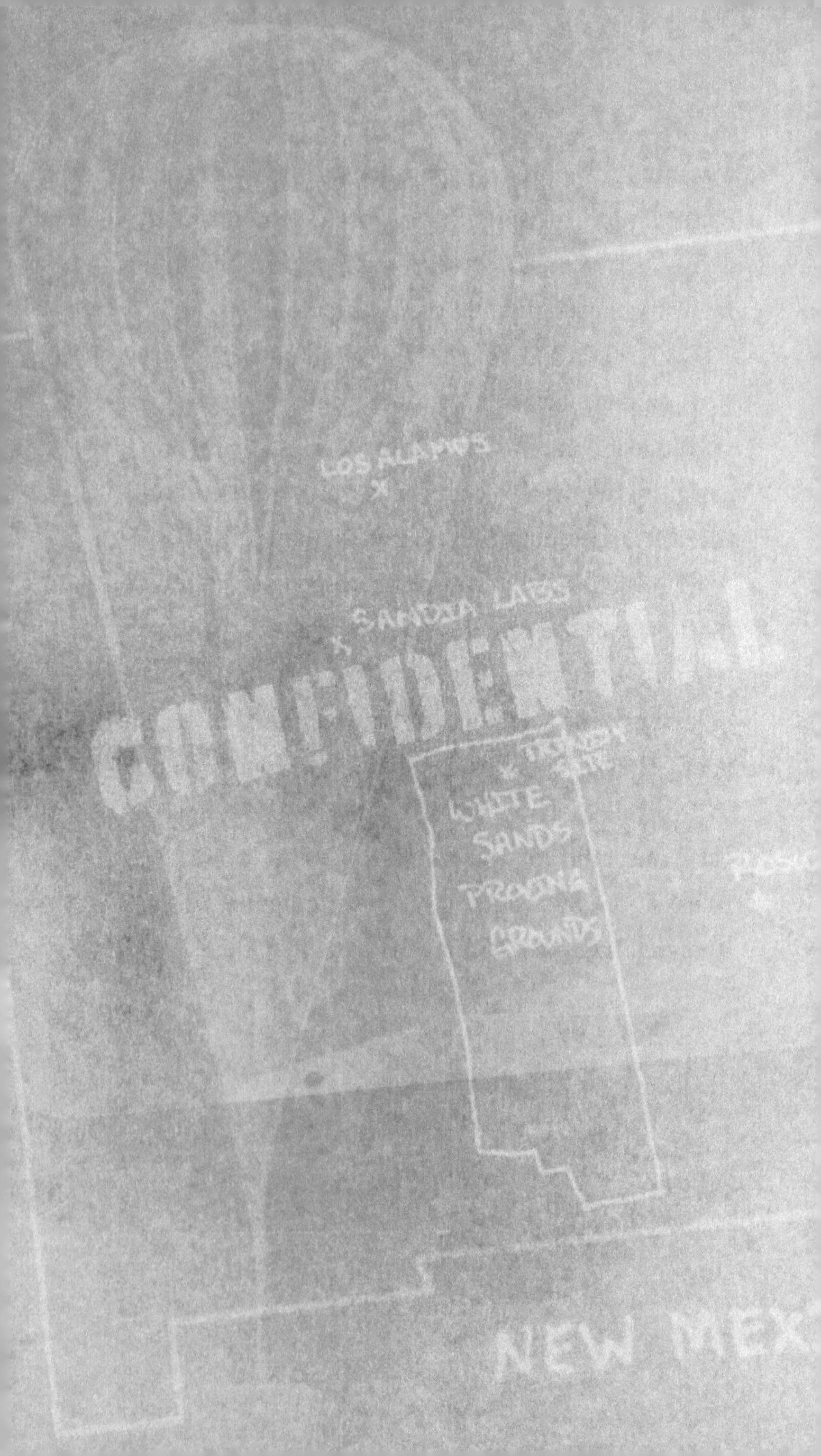

LOS ALAMOS
X
X SANDIA LABS
CONFIDENTIAL
X TRINITY SITE
WHITE SANDS PROVING GROUNDS
NEW MEX

CHAPTER 2
"Just the Facts, Ma'am"

AS SERGEANT JOE FRIDAY used to say in the classic television crime drama *Dragnet*, "Just the facts, ma'am." Actually, Sergeant Friday never said it precisely this way, and if you thought you knew the phrase was verbatim from actor Jack Webb's lips, perhaps you should read carefully from here on. Many things are not exactly as they seem.

The US government, our government, through the US Army Air Forces and the US Air Force told us four Roswell stories over the past seventy-five years. Simple, everyday logic dictates clearly that at least three are false.

- The first story is the original press release directed by Colonel William Hugh "Butch" Blanchard, Commanding Officer, 509th Bomb

Group and Roswell Army Air Field, reporting the capture of a flying saucer or disk. The article appeared initially in the *Roswell Daily Record's* evening newspaper on July 8, 1947. The public information officer of the 509th at that time, First Lieutenant Walter Haut, is commonly associated with the press release, but there should be no speculation about its true author. It was ordered by the Commanding Officer, Colonel Blanchard.

- The second story is the weather balloon story, told by Blanchard's boss, Brigadier General Roger Ramey, Commanding Officer, 8th Air Force, Fort Worth, Texas, less than twenty-four hours later, shortly after Colonel Blanchard had conveniently gone on leave. Colonel Blanchard never appears as part of any Air Force story about Roswell afterward.

- The third story is told in the three US Air Force Roswell reports of 1994-97 led by United States Air Force counterintelligence guru Colonel Richard L. Weaver. This includes the 1997 US Air Force report that throws anthropometric dummies into the mix. The

1994 report is the top-secret Project Mogul story of almost 1,000 pages.

- The fourth story is the non-story of omission. It began just after the second story – the weather balloon story – and continues to this day. Other than three or four exceptions (for example, the *Case Closed* report), the United States Air Force started not talking about the Roswell Incident on July 9, 1947, and continues to not talk about Roswell. In fact, the Incident is conspicuously absent in the 2021 Department of Defense report on Unidentified Aerial Phenomena.

At least three of these stories are not true. The second, third, and fourth stories are, effectively, rewrites of the initial report – the original press release and subsequent newspaper article on the Incident. The rewrite, which I witnessed during the Walker Air Force Base Museum's opening ceremony on September 18, 2010, is part of the rewrite of omission. And again in 2021, when the Pentagon issued an update on UFOs, changing the nomenclature to UAP (Unidentified Aerial Phenomena), the US Air Force still isn't talking about Roswell.

The United States Air Force has a pretty good record of keeping secrets, and everything they communicate about Roswell has been and continues to be about keeping secrets.

1947 – SETTING THE STAGE

The modern history of Unidentified Flying Objects or Unidentified Aerial Phenomena effectively began on June 24, 1947, when private pilot Kenneth Arnold reported seeing "flying saucers" in flight near Mount Rainier in Washington state. The best summary article I have read is in *The Atlantic Magazine's* issue dated June 15, 2014. It was written by Megan Garber and covers most of the facts in an easy-to-read, fact-based manner. Arnold's report of what he witnessed and the media coverage that resulted sparked a broad-based public response. The UFO phenomenon was off to the races.

Arnold did not specifically report seeing "flying saucers." He described crescent-shaped aircraft flying in formation.

The United States Air Force (United States Army Air Forces at the time) was careful and reticent in its public responses. Much of the Air Force response history is hard to document, but two factors can be identified:

1. The Air Force didn't have an official, public response.

2. The Air Force had concerns about protecting the airspace of the United States against possible enemies who could fly faster, turn quicker, and operate without detection. In intelligence this is called a *vulnerability*. Vulnerabilities are typically the most classified of all defense information.

The Air Force should not have been surprised by Arnold's or Blanchard's 1947 reports of so-called flying saucers. During World War II, military pilots on all sides and in almost every theater of the war reported strange, unusual, and oftentimes scary flying objects they observed. The term *foo fighters* came into use as the sightings mounted, new enemy weapons were encountered, and the capabilities of all weapons expanded. Rumor has it that the US Army Air Forces followed up on the reports after the war, but somehow their research "got lost."

Some years after the Roswell Incident, the Central Intelligence Agency (CIA) got a group together to investigate. In 1953, California Institute of Technology physicist Howard P. Robertson chaired a panel of six scientists with expertise in experimental aviation technology to help the CIA determine if foo fighters constituted a national security threat. The panel seemingly reached no official conclusion, as reported in the August 2016 *Air & Space Magazine* article, "What Were the Mysterious 'Foo Fighters' Sighted by WWII Night Flyers?"

The UFO mission was almost certainly a low priority in the first half of 1947 for the United States Army Air Forces. At least three other mission areas were most likely of greater concern:

1. Nuclear warfare;

2. Emerging efforts in rockets, radar, and the high ground of space; and

3. Preparations to become a separate, equal service by deleting the "Army" in the organizational name, scheduled for September 1947.

Of course, just as the city of Albuquerque is a "star" of the television series *Breaking Bad*, the state of New Mexico must be recognized as a key element in all these Air Force mission priorities. It doesn't take a rocket scientist to hypothesize that if extraterrestrial intelligence chose to visit planet Earth in 1947, they might be drawn to New Mexico by what was going on in the high desert.

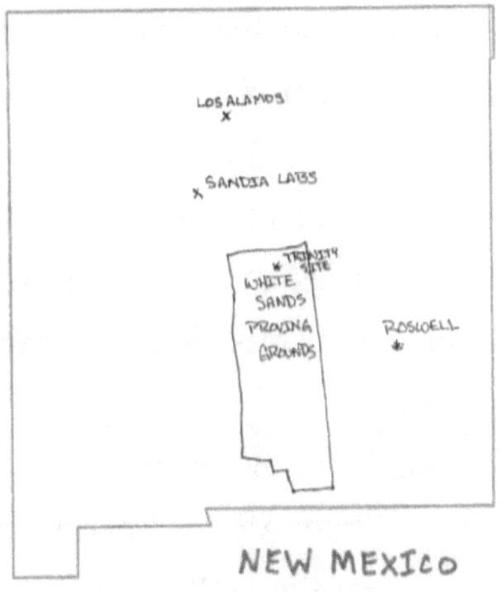

Author's hand drawn map of New Mexico noting US Military Research oriented 1947 sites

PRIORITY 1 – NUCLEAR WARFARE

July 16, 1945. Trinity Site, White Sands Proving Ground, New Mexico. The atomic age began with the first explosion of an atomic bomb. J. Robert Oppenheimer, Major General Leslie Groves, Enrico Fermi, and other physicists, scientists, and military officials from the Manhattan Project were on site at White Sands Proving Ground, having driven south from Los Alamos, New Mexico, where they had built the gadget. (As an aside, local legend has it that they stopped at the Owl Bar in San Antonio, New Mexico, where they had some part in the invention of one of New Mexico's greatest cultural and gastronomic contributions, the green chile cheeseburger.) You can visit the historical Trinity Site and marker in person twice a year when the US Army hosts an open house event for the public, but normally the classified activities at what is now called the White Sands Missile Range prevent public access to the site.

In 1947 the McMahon Act went into effect, establishing the United States Atomic Energy Commission to effectively replace the WWII Manhattan Project and expand its mission. Los Alamos, New Mexico, which began as one of the key Manhattan Project sites of WWII, continues to be associated with nuclear research. Kirkland Air Force Base in Albuquerque, the current home of Sandia Laboratories, has a history of integration with the Manhattan Project, and in early 1947 received a new mission from Air Materiel Command of the US Army Air Forces for nuclear weapons testing and evaluation.

The 509th Composite Group, which was the only military unit in the world with nuclear weapons capability in 1945, evolved after WWII to become the 509th Bomb Wing and found a post-war home at Roswell Army Air Field in Roswell, New Mexico. In 1946 the 509th conducted atomic operations in support of Operation Crossroads research at Kwajalein Atoll in the Pacific. From 1945 to 1947, New Mexico was a hotbed of classified, atomic weapons activity, much of which continues today.

PRIORITY 2 – EMERGING EFFORTS IN ROCKETS, RADAR, AND THE HIGH GROUND OF SPACE

The first successful effort of exploration in space occurred at White Sands Proving Ground in New Mexico on May 10, 1946. Yup, you read that right: 1946 in New Mexico. Not the Russians' launch of *Sputnik* in 1957.

Operation Paperclip, the capture of Nazi scientists involved in the V-2 rocket program during WWII, resulted in almost 300 V-2 rockets and parts being captured by the United States Army along with the scientists. When the US Army needed a place to put the rockets and scientists, they picked White Sands Proving Ground (later White Sands Missile Range). It had already been chosen a year or two earlier for another successful scientific milestone: the first explosion of an atomic bomb.

Around fifteen of the V-2s were launched in 1947 including one launched on May 29 that went off the range and landed near a cemetery in Ciudad Juarez, Mexico. Certainly, this one didn't meet any secrecy/classification requirements when it crossed into a foreign country and showed up in the media.

V-2 launches continued through 1952. Despite the reference staff on the television show *Jeopardy* occasionally answering to the contrary, the first living things – including mammals – were launched from White Sands Proving Ground. The first photographic evidence of the curvature of the Earth was accomplished on one of these early missions. There is a museum that documents most of these facts just inside the US Highway 70/Owen Road entrance to the main post area that welcomes public visitors to White Sands. A separate building across the parking lot from the museum proper houses a rebuilt V-2.

Space. The final frontier. White Sands Missile Range (WSMR), Roswell, and New Mexico have been associated with rocket science and space exploration for a long time, and some of the associations may come as a surprise to you.

Analogous to those Wright boys from Ohio heading to the Outer Banks of North Carolina to test their flying machine, Dr. Robert H. Goddard left the East Coast to find fertile rocket launch fields in Roswell, New Mexico in the summer of 1930. There are several museums in Roswell. Dr. Goddard is not in the UFO museum. He has an entire wing of the larger Roswell Museum building named for him as well as a high school. In an interesting trivial observation, one of Dr. Goddard's early rockets was designated A-4, the same designation used by the Nazis for their original V-2. Dr. Goddard's last launch in Roswell was May 1941, less than a year before the United States' entry into WWII.

Radar and stealth technologies have been developed at White Sands Missile Range since its inception, but the public is not generally aware of the classified activities in these areas. If you visit White Sands National Park, which is entirely within the confines of WSMR and Holloman Air Force Base, you should climb up on a tall dune and look to the west-northwest – you might be able to see a distinctive slash in the desert landscape at the foot of the San Andres Mountains. This site, along with some "special" towers (slanted, white, wooden, tall, thin), are radar-related sites. Stealth technology research began there and most likely continues to this day.

In summer 1992 the F-117 Nighthawk fighter force was moved from Tonopah, Nevada, to Holloman Air Force Base outside of Alamogordo, New Mexico. Another US Air Force counterintelligence success – keeping secret the existence of these fighters – was successful for at least the first six years after they began flying. In some ways, the move to New Mexico might have been seen as a "homecoming" of the technological advances that most likely were initiated at the White Sands Missile Range in New Mexico's Tularosa Basin, a 3,200-square-mile field of restricted desert and mountains that includes Holloman AFB.

In addition to missile firing exhaust trails, spacecraft landings, and other visible phenomena, there is plenty of activity that is invisible. The Area Frequency Coordinator is a busy office at White Sands Missile Range. This entity is responsible for ensuring that the electromagnetic frequency spectrum is controlled to prevent accidents, interferences, successful intelligence activities by adversaries, and other problems that might arise from different frequencies being used by different folks at the same time or location. While the activity is invisible to the human eye, almost all the electromagnetic spectrum is probably represented in the air above the high desert terrain.

Priority 3 – Becoming a Separate Service

The US Army Air Forces became the independent US Air Force on September 18, 1947, when the 1947 National Security Act went into effect, reorganizing the defense and intelligence agencies of the United States post-WWII. Stuart Symington was appointed the first Secretary of the Air Force, and General Carl A. Spaatz became the first US Air Force Chief of Staff. Colonel William H. "Butch" Blanchard was Commanding Officer, 509th Bomb Group in Roswell (formerly the 509th Composite Group of WWII), and Brigadier General Roger Ramey was his Commander at 8th Air Force Headquarters of the new Strategic Air Command in Fort Worth, Texas.

Even after the Roswell excitement back in July 1947, all seemed to be well and right in the new US Air Force by September. The "weather balloon" story was holding up. The monopoly of nuclear warfare was holding up. The vulnerabilities of the United States to unknown aerial threats were still being debated within the new service and any related information was still classified.

It was in this historical setting – in the power of place state of New Mexico and with the soon-to-be independent service – that something dropped out of the sky and our four stories begin. As you read these detailed accounts, keep in mind that – as stated earlier – logic dictates that at least three versions must be false. Also keep in mind that the US Air Force has openly admitted they were not truthful about the initial weather balloon story, later identified as the Project Mogul data-gathering balloon.

STORY #1: THE PRECIPITATING EVENT AND THE ORIGINAL PRESS RELEASE – COLONEL WILLIAM H. BLANCHARD

It doesn't rain much in New Mexico, but when it does, more often than not, it is a thunderstorm. Such was the night of July 2, 1947, when a massive thunderstorm struck Roswell, New Mexico and the Roswell Army Air Field. From the third of July until the eighth, life around Roswell appeared to go on pretty much as normal. It is generally hypothesized that the thunderstorm on July 2 might have had something to do with the Incident.

It is also generally accepted that the Incident began on July 8, 1947, with the *Roswell Daily Record*, an evening paper, reporting that Roswell Army Air Field Commander Colonel "Butch" Blanchard and his troops had captured a flying saucer. The Associated Press picked up the evening story and set off a global firestorm.

While the article in the *Roswell Daily Record* newspaper was probably just the tip of the iceberg, it was effectively chilled the very next morning by the weather balloon story out of Fort Worth, Texas, when Colonel Blanchard's superior, Brigadier General Roger Ramey of the 8th Air Force threw cold water on the Roswell flying saucer. It was a weather balloon, he said, and he had pictures of it taken right there in his office.

It is upon this precipitating event that all the rest of the stories, books, testimonies, official government reports, alien autopsy videos, television shows, movies, and serious economic development in the city of Roswell rest. The Air Force's current explanation is that a mistake in identification was the cause of the hullabaloo.

There is general agreement about the core details of the original reporting:

- Ranch foreman Mac Brazel from the Foster Ranch near Corona, New Mexico, northwest of Roswell, reported finding debris on the ranch to the Chaves County sheriff and subsequently to the Roswell Army Air Field.

- Major Jesse Marcel, 509th Bomb Group Intelligence Officer and Captain Sheridan

Cavitt, Counterintelligence Officer (not under 509th/Colonel Blanchard command) were ordered to the debris site and collected debris. Major Marcel stopped at his home on the way back to his duty station and shared a review of some of the materials recovered with his wife and his son, Jesse Marcel, Jr.

- Colonel Blanchard ordered a press release be issued through his public affairs officer to local radio stations and the *Roswell Daily Record* newspaper on July 8, 1947, reporting the capture of a flying saucer.

- All hell broke loose, and the world became acutely aware of the 509th, Roswell Army Air Field, UFOs, New Mexico, and the city of Roswell for a short while, about twelve to eighteen hours.

- Brigadier General Roger Ramey, Blanchard's Commander at 8th Air Force Headquarters in Fort Worth, Texas, issued a press release the next day (less than twenty-four hours later) – July 9, 1947 – that changed the original report to a story about a weather balloon.

Why did Colonel Butch Blanchard issue the original release? What the heck was he thinking? How could he and his men have made such a colossal mistake in identification between a weather balloon and a flying saucer? Remember, this group regularly handled state-of-the-art military aircraft and highly technical nuclear weapons.

Did the man who later died of a massive heart attack while on duty at the Pentagon as General Blanchard (four stars), Assistant Chief of Staff of the USAF – one step away from the sure bet of becoming Chief of Staff, the highest-ranking USAF officer – have a wicked sense of humor? Was it an inadvertent disclosure (a counterintelligence or CI term)? Was he telling the truth at the time, only to find out later that he had no need to know (another CI term)? Why were the two officers from Intelligence and Counterintelligence (Major Marcel and Captain Cavitt) dispatched to check on the site? Did Blanchard actually go to the site to see the wreckage for himself? Or did he rely on others' reports of what they found?

There are hundreds of unanswered questions. While many have obvious answers, some are true mysteries. And, of course, there are all those personal stories presented in myriad books, articles, and online accounts as well as in-person discussions in Roswell most years at the annual UFO Festival by an army of Believers.

The press release is a fact. Subsequent tales and "testimony" about alien bodies, the spacecraft itself, transfer of materials to other locations, additional crash sites, and the like are not acknowledged by the entire (or almost the entire) Believers and Debunkers communities. The press release and *Roswell Daily Record* newspaper article have not generally been contested. They are fact. In spite of some of the Air Forces' best CI efforts, there can be little question that Colonel Butch Blanchard issued the press release, and he wasn't joking.

STORY #2: THE WEATHER BALLOON – BRIGADIER GENERAL ROGER RAMEY

Within twelve hours of the Incident's initial press release, Colonel Butch Blanchard's role in the Incident had been superseded by his superior officer, Brigadier General Roger Ramey, Commander, 8th Air Force at Fort Worth Army Air Field, Texas. In fact, Colonel Blanchard immediately had his leave approved and went on leave from Roswell. He doesn't really show up again in any Air Force UFO documents to this day.

Brigadier General Ramey ordered Blanchard's 509th Intelligence Officer Major Jesse Marcel to fly to Fort Worth Army Air Field, and they both had their pictures taken with weather balloon debris in front of a number of newspaper reporters and photographers. This is a fact. There is disagreement about what exactly was photographed, who was doing what, and what that photographed telegram in General Ramey's hand said. Yet few disagree about these two facts: the US Army Air Forces issued a second press release, and this changed the story. The weather balloon story is a fact.

Why did General Ramey act so quickly? Why were the photos taken at Fort Worth Army Air Field instead of Roswell Army Air Field? Why didn't Counterintelligence Officer Captain Sheridan Cavitt fly to Fort Worth? What *did* that telegram in General Ramey's hand say? Are we really expected to believe that identification of balloon debris is only possible by generals – not colonels, majors, or lieutenants? Why did the Air Force choose General Ramey to become the point man in the service's effort to discount the Roswell Incident and, essentially, all UFOs? And where the heck was Lieutenant Colonel Payne Jennings, who assumed command of the 509th when Colonel Blanchard went on leave?

Story #3: Project Mogul – Colonel Richard L. Weaver

In 1994, Congressman Steven Schiff of New Mexico's 1st US Congressional District in northern New Mexico asked the US Government Accounting Office to investigate the Roswell Incident. The GAO went looking for help in the investigation, and the US Air Force "volunteered" to assist in looking through its records. The Air Force and the Government Accounting Office ended up providing four separate documents (two large reports and two smaller documents) in 1994, 1995, and 1997:

- *Report of Air Force Research Regarding the "Roswell Incident," with Memorandum for Secretary of the Air Force*, by Richard L. Weaver, Col., USAF, July 27, 1994. (Note: This document appears to be included completely as part of the *Fact versus Fiction* document released in 1995.)

- *The Roswell Report: Fact versus Fiction in the New Mexico Desert*, Headquarters United States Air Force, Colonel Richard L. Weaver and 1st Lieutenant James McAndrew, 1995.

- *Results of a Search for Records Concerning the 1947 Crash Near Roswell, New Mexico,* Government Accounting Office, July 1995.

- *The Roswell Report: Case Closed*, Captain James McAndrew, 1997.

Why was the US Air Force's chief counterintelligence (CI) person, Colonel Richard L. Weaver, assigned to lead the effort to gather the documents, compile the evidence, and write these reports? Why not a public affairs officer, the US Air Force's chief historian, US Air Force attorneys, or even some scientific types assigned to the job? Why the Air Force? Is the Department of Defense not cleared by the USAF for access to Roswell secrets? Project Mogul was a great explanation in the *The Roswell Report: Fact versus Fiction in the New Mexico Desert*. Does it seem that it was a little too convenient? *Fact versus Fiction* does provide us with a look into the specifics of the investigation of the Air Force by the Air Force. And the transcript of the interview conducted by Counterintelligence Officer Colonel Weaver of retired Air Force officer Sheridan Cavitt, one of the few people to have been an eyewitness to the flying saucer/Project

Mogul balloon, is very telling. Telling meaning that it creates more questions than answers.

Are we really expected to believe the "add on" report of 1997, *The Roswell Report: Case Closed*? It seems to have been written to specifically counter conspiracy theories about alien bodies and autopsies. Can we really believe accounts of test dummies from five to ten years later to explain stories of alien bodies from 1947? Why should the original story include bodies anyway? Colonel Blanchard's press release did not mention bodies.

STORY #4: OMISSION

From July 9, 1947, to the present day, the final attempt to rewrite the history of the Roswell Incident includes the rewrite of omission.

For many, 1947 was the year that modern-day UFO phenomena really began. As mentioned earlier, in June of that year private pilot Kenneth Arnold reported sighting a formation of UFOs near Mount Rainier in the state of Washington. The *Chicago Sun-Times* newspaper coined the term *flying saucer*, and a new cottage industry, including a publishing category, thank you very much, was born.

By the time the Roswell Incident occurred in July 1947, the Army Air Forces had assumed the UFO leadership role for the US government, determining that Unidentified Flying Objects in US airspace might be a threat. It was a time of uncertainty in US and world history. World War II was over, the Cold War was just beginning in earnest, and the Army Air Forces was in the process of becoming a separate but most self-important armed service. The Army Air Forces, the world's only combat-proven nuclear force at the time, was preparing for the future of warfare that placed them at the point of the spear.

Southern New Mexico was a hotbed of US government confidential, secret, and top-secret activity. At White Sands Proving Ground, some 125 miles west of Roswell and the 509th Bomb Group, captured and rebuilt German V-2 rockets were being launched in the very first explorations of space by teams of government, military, contractor, and former Nazi scientists. The very first successful rocket launch in the exploration of space by humans took place from Launch Complex 33 on May 10, 1946. (Launch Complex 33 is now a National Historic Landmark.) The first launches of living things, live animals, took place there too. All of this in secret more than eleven years before the Russians launched *Sputnik* in 1957.

The White Sands Proving Ground evolved into the White Sands Missile Range, and lots of classified activity still goes on there. Every April and October the general public has access to the Trinity Site where the world's first atomic device was exploded in 1945, almost two years to the day before the Roswell Incident.

THE AIR FORCE INVESTIGATION PROJECTS, 1947-1997

Over the years, the US Air Force has investigated itself a number of times regarding UFO/UAP. These investigative efforts included the well-known Project Blue Book. However, none of the efforts prior to 1994 included Roswell. None of the efforts included Roswell. *For the final time: None of the efforts included Roswell.*

Investigation #1. Project Sign, 1947-1949 — With the publication of the *CONFIDENTIAL Classified UFO Subject: Project Sign* report, the Incident at Roswell had already disappeared. The listing of UFO reports includes plenty of 1947 records but no mention of the Roswell Incident. Interestingly, the files in the same Box 3678 of the National Archives Record Group 342, Stack 190, Row 72, Compartment 27, Shelf 3-5 include some photographs of flying wing-type aircraft. A copy of the Project Sign report is also on the Defense Technical Information Center website, dated February 1949 and originally classified as SECRET. (https://apps.dtic.mil/sti/pdfs/AD0311102.pdf).

Investigation #2. Project Grudge, 1949-1951 – This UFO follow-up effort by the US Air Force to Project Sign did not include Roswell.

Investigation #3. Project Blue Book, 1947-1969 – The United States Air Force mother-of-all-UFO-research efforts and successor to Project Grudge does not include Roswell.

Investigation #4. Condon Report, 1968 – The published report, titled *Scientific Study of Unidentified Flying Objects,* was a contract report by the University of Colorado at Boulder, which became known simply as the Condon Report. The history of the contracting, supervision, and writing of this report most likely had a great influence on scientists and scientist "wanna-be's." And that influence would have multiple results. First, it began with the scientific community already divided along the Believer-Debunker continuum and did little or nothing to move folks on this continuum, "scientifically" or otherwise. Second, it was initiated, sponsored, funded, and controlled by the US Air Force. Third, it continued the omission of Roswell.

Finally, its conclusions stand in stark contrast to what science and scientists have pursued since the mid- to late-1960s: finding new planets capable of life as we know it, developing new devices to explore the cosmos looking for evidence of life, and expanding scientific efforts to explain historical phenomena. For decades, we have been making technological advances specifically designed to search for evidence of extraterrestrial life. In 2022 the James Webb Space Telescope reached its Lagrange Point L2 stationary point some one million miles from Earth where it will look for new planets and attempt to locate the edge of the universe. A robot helicopter on Mars is looking for signs of life, current or historical. NASA intermittently, almost weekly, reports the discovery of a new planet capable of life as we know it, an exoplanet, also dubbed Goldilocks or Cinderella planets. The US Navy has gone public with Naval Aviators' sightings of UFO/UAP.

But for this author, the single biggest result of the Condon Report was documenting and providing evidence of the continuing efforts of the United States Air Force counterintelligence campaign.

The history of the Condon Report (three Volumes) is worthy of your interest, and you can find it online at one of my favorite sources, The Defense Technical Information Center (DTIC). If you do a search of authors, you may find me in there somewhere.

Investigation #5, 1994 – Finally, the United States Air Force had no choice and was unable to avoid direct accountability for the Roswell Incident. Congress came knocking at the door, and the Air Force had to answer. This resulted in the publication of the previously mentioned government report, *The Roswell Report: Fact versus Fiction in the New Mexico Desert*, by Colonel Weaver and 1st Lieutenant James McAndrew, released in 1995.

In this report, Project Mogul, a bigger and classified data-gathering "weather balloon" project was offered up as explanation. The nearly 1,000-page report had plenty of photos of balloons, page after page of meticulously hand-recorded data, and the transcript of an interview with Counterintelligence Officer Lieutenant Colonel (Retired) Sheridan Cavitt (Captain in 1947) "... the only surviving person who recovered debris from the original Roswell site in 1947." By using this wording, the Air Force avoids talking about Colonel (Dr.) Jesse Marcel, Jr., who consistently claimed to have handled some of the debris collected by his father, Major Jesse Marcel, when he was a boy. It was the complete reading of this 1995 document that prompted your author's determination to conduct extensive research and put this book into writing.

Three specific parts of the 1995 *Fact versus Fiction* report Introduction call for immediate attention.

First, this phrase on page one of the Introduction: "... the facts regarding the reported crash of an UFO in 1949 [1947] at Roswell ...." Did the General Accounting Office or the US Air Force ensure that this confusion about the actual year of the Roswell Incident was documented?

Second, again on page one of the Introduction: "In short, the objective was to tell the Congress, and the American People, *everything* the Air Force knew about the Roswell claims." The italics in this quote are by the Air Force, not your author. Instead of completely discussing the Roswell "Incident" the Air Force chooses to only look at the Roswell "claims"?

Lastly, when first reading the transcript of Colonel Richard Weaver's interview of Lieutenant Colonel (Retired) Sheridan Cavitt (Counterintelligence Captain in 1947) and his wife, it appeared that the "interview" might not be completely free of predisposition or bias on everyone's part. My first reading of the transcript motivated a trip to the National Archives and Records Administration facility in College Park, Maryland, to listen to the actual taped interview. It was well worth the effort.

My initial impressions of a predetermined outcome of the interview supporting a balloon conclusion from the transcript were validated when I listened to the recorded interview. During the interview, Colonel Weaver routinely states the answer he is looking for as a question, and Lieutenant Colonel (Retired) Cavitt replies affirmatively most of the time, but he often claims memory loss. Cavitt is often assisted by his wife in his answers. There is a gap in the taping for an apparent technical problem. You are strongly invited and encouraged to follow in my footsteps, listen to the actual interview, and make your own decision.

As I noted in Chapter 1, in 2010 at the dedication of the Roswell Army Air Field/Walker Air Force Base Museum, the Roswell Incident was still missing from the ceremony, speeches, and museum exhibits. The counterintelligence campaign of omission continues, even decades after the Incident.

There are literally thousands of books about the Roswell Incident that have been written and published by other people and organizations – not the US Air Force. It appears that for about every 100 books on the Believer side of the continuum, there are only a couple on the Debunker side. The International UFO Museum and Research Center in Roswell, New Mexico, seems to have most of them (both categories) on the shelves of their research center as well as a multitude of filing cabinets containing other written materials.

You can quickly find yourself on the Debunker side of the continuum simply because of the intricate and extensive maze of stories that have been generated since the original event. As one Debunker book points out, an entire mythology surrounding Roswell has grown during the seventy-five years since the Incident.

This book is focused on the facts associated with the original Incident, the subsequent US Air Force activities regarding the Incident, and the subsequent omissions.

THE KISS PRINCIPLE

Most sources give famed aerospace engineer Kelly Johnson of the Lockheed Skunk Works credit for the concept of KISS – Keep It Simple, Stupid. There are a few folks, probably Navy veterans, who want to blame the Navy, but no matter who created the concept, it is applied here to the Roswell Incident. In engineering and science things can sometimes become complicated and counterintuitive.

When you apply the KISS principle, you can ask yourself this question: Could the 509th Commander and his team not identify a weather balloon?

LOS ALAMOS
X
X SANDIA LABS
CONFIDENTIAL
TRINITY SITE
X
WHITE
SANDS
PROVING
GROUNDS
NEW MEX

CHAPTER 3
Back to the Beginning

THE BEGINNING was the press release ordered and authorized by the Commanding Officer of the 509th Bomb Group at Roswell Army Air Field, Colonel William H. "Butch" Blanchard and published in the *Roswell Daily Record* on Tuesday July 8, 1947.

Again and again this initial event looms large and is still, seventy-five years later, largely unexplained. The unanswered questions are myriad and critical:

Did Colonel Butch Blanchard perpetrate this press release as a practical joke, only to be called on the carpet within twelve hours by his superior, Brigadier General Roger Ramey?

Did First Lieutenant Walter Haut, the 509th Public Affairs Officer act alone and without Colonel Blanchard's approval in handling the release?

Why were the first two Air Force personnel to investigate from intelligence and counterintelligence?

Why did the Incident get bumped up to higher headquarters so quickly?

Why would Colonel Jesse Marcel, Jr., son of 509th Intelligence Officer Major Jesse Marcel, stick to his story until his death about the debris he handled with his dad, including attempts to burn it and cut it? Both Major Marcel and his son always referred to the material as "not of this world."

Why do so many people in Roswell, civilians and former military personnel alike, still have doubts about the Air Force's handling of the initial event?

Was that first press release really presenting the truth?

And what was Charles Lindbergh doing in New Mexico in 1947?

Charles Lindbergh? Sorry, just a quick insert to demonstrate one of the distractions used in this counterintelligence campaign and to demonstrate how there are literally thousands of potential witnesses, third-party testifiers, and ancillary documents at use in this CI campaign. Roswell Believers are sometimes their own worst enemies in their efforts to penetrate the US Air Force's CI campaign to protect the information of 1947. They are doing the Air Force's work for them when they play the "he said/she said" game and base their conclusions on hearsay "witness" testimony. Lindbergh was in New Mexico in 1947, but this probably had nothing to do with the Roswell Incident.

From the very beginning there was some division in the US population's beliefs about Roswell. Back then, in 1947, most folks generally believed what their government told them and ended up initially on the Debunker side of the Believer-Debunker continuum. This might be one of the reasons the secrets were generally held close for over thirty years. Any consideration of extraterrestrial intelligence was more likely to be in the science fiction arena, not the sciences. It would essentially remain there until the beginning of the 1960s when the first search for extraterrestrial intelligence (SETI) efforts began.

In 1978 books about Roswell began to appear, and the game was afoot. Believers' numbers began to swell almost overnight, and next thing you knew there was an annual celebration established in Roswell to capitalize on its sudden notoriety. By the 1990s you could identify yourself as being from New Mexico to foreigners, and they would only know the state from the mention of that widely acclaimed UFO crash site, not Santa Fe or Albuquerque or even

Alamogordo and the first atomic bomb. For a while some of the locals even reported that you were guaranteed financial wellbeing by moving to Roswell and opening a T-shirt shop.

Believer writers such as Stanton Friedman, William Moore, Kevin Randle, and Don Schmitt began retelling the stories they heard from 1947 residents, and suddenly there was a labyrinth of plots, subplots, characters, and the addition of alien bodies.

Today, one can get misdirected, misinformed, deluded, and otherwise distracted. As mentioned above, some Believers are contributing to the Air Force's counterintelligence campaign by creating a confusing morass of fantastical stories that are not credible, logical, or believable. This book will attempt to stick to facts, simple-to-understand reasoning, and evidence that can be experienced directly by the reader.

Keep it simple, stupid: KISS. Whether Kelly Johnson, famous Lockheed Skunk Works aircraft designer, or the US Navy invented the phrase, it has become a staple in engineering endeavors and other technical fields. It is the antithesis of the CI campaign, and it is the guiding principle of this book.

Chapters 6, 7, and 8 of this book focus on the three key, critical players in this story. There are enough players in the broader story that if numbers and words could swing opinions this tome might include thousands of pages of "he said/she said," third-party, and hearsay "evidence." The US Air Force *Fact versus Fiction* report runs almost 1,000 pages with plenty of pictures of balloons, lots of characters, and almost endless data sheets. Perhaps volume shouldn't be the determinant of truth.

During the almost eight years I lived in Roswell, there was always an individual story recounted from 1947 residents when I asked about the Incident. Most of the stories were short of facts but long on opinion and speculation. My impression was that approximately two-thirds of those who talked about the Incident with me were on the Believer end of the continuum. Some of the locals' stories have grown to further the mythological aspects of the Incident, including reports of child coffins ordered and used for deceased small-bodied aliens. We will not go there.

The three individuals included here were carefully selected. All three are military men; all three were colonels. Two of them, Colonel Blanchard and Colonel Weaver have "a dog in the fight," representing in many ways the opposite ends of the case. Colonel Blanchard issued the "flying saucer" press release in 1947, while Colonel Weaver authored *The Roswell Report: Fact versus Fiction in the New Mexico Desert* in 1995.

The third individual was a bystander, eyewitness, and seemingly impartial participant who was persuaded by direct, personal experiences to the Believer side of the continuum. Because of his experiences, history, accomplishments, and lifelong consistency he stands out as credible, truthful, and believable. He was not interviewed in the 1990s for the Air Force's *Fact versus Fiction* and *Case Closed* reports. Wonder why they didn't talk to him? Colonel (Dr.) Jesse Marcel, Jr. and his wife wrote their own book, *The Roswell Legacy: The Untold Story of the First Military Officer at the 1947 Crash Site*, in 2008.

In my research for this book, I maintained a spreadsheet of Incident People that eventually grew to almost a thousand individuals. There were more than fifty in the spreadsheet assigned to the 509th Composite Group (later named the 509th Bomb Wing) in 1947. In the end, the three individuals presented here were judged the most important to helping us understand the Incident.

A side note: Reviewing the 1947 509th unit yearbook at the UFO Museum in Roswell verified that segregation was still in effect as pictures from the

"Service Company" showed White faces in command positions and Black faces throughout the rest of the unit. Access to materials like this unit yearbook are one reason to visit the UFO Museum, no matter which side of the Believer-Debunker continuum you fall on.

During my residence in Roswell, I attended the annual UFO Festival on numerous occasions. Lots of authors on both sides of the Believer-Debunker continuum attended these events, and I met most of them. At one of the events, I met Stanton Friedman who is generally credited with putting Roswell back in the public view in the late 1970s and early '80s. We traded phone numbers, and once or twice while I was at the National Archives and Records Administration facility in College Park, Maryland, he took my phone calls and helped me locate records. He was always friendly, unobtrusive, and grounded in his discussions about the Incident.

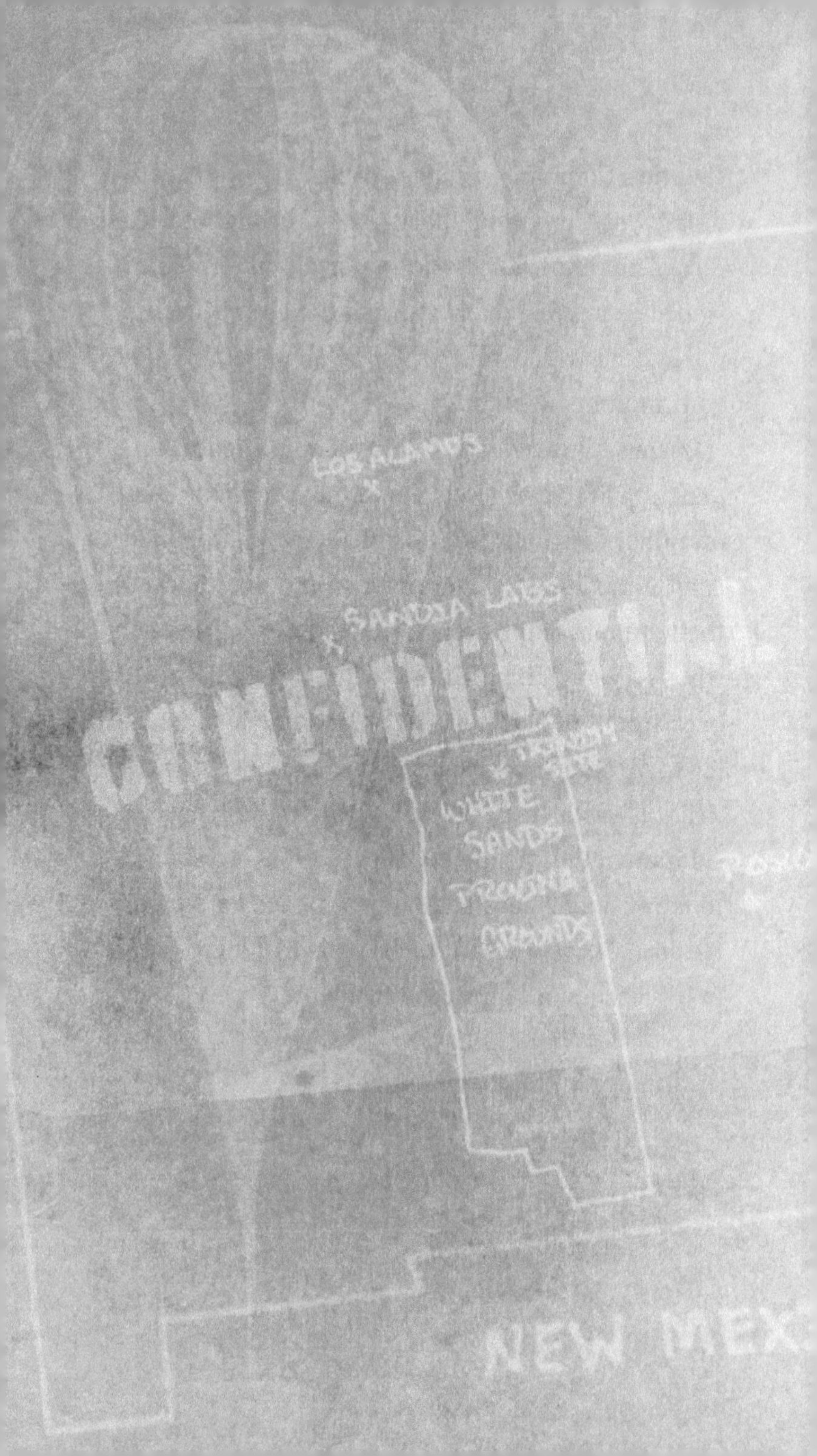

LOS ALAMOS
SANDIA LABS
CONFIDENTIAL
WHITE SANDS PROVING GROUNDS
NEW MEX

CHAPTER 4
Counter-intelligence

THE COUNTERINTELLIGENCE (CI) campaign began in 1947 and continues to this day. Air Force intelligence and counterintelligence have been a part of the Roswell Incident from the very beginning through the "case closed" Air Force-defined end in 1997. Today, the counterintelligence or CI campaign by the US Air Force to protect information from 1947 continues.

Coincidentally, 1947 was a milestone year for Roswell, the intelligence communities, the US Air Force, and some fellow named Chuck Yeager:

- July 8, 1947 was the beginning of the Roswell Incident.

- July 26, 1947 saw the National Security Act of 1947 go into effect.

- September 18, 1947 was the birth of the new, independent military service called the United States Air Force, mostly the former US Army Air Forces.

- October 14, 1947 saw USAF Captain Yeager break the sound barrier.

Since intelligence and counterintelligence are integral parts of the Roswell story, it is critical to consider their implications in all aspects of the Roswell Incident.

There is lots of information on the Internet about intelligence and counterintelligence both in general and as they relate to the Department of Defense (DoD) and the United States Air Force. One of many useful documents to understand the history of the US intelligence and counterintelligence organizations from 1947 to the present is "An Assessment of the Evolution and Oversight of Defense Counterintelligence Activities," written by Michael J. Woods and William King, published in the *Journal of National Security Law & Policy*, Vol. 3: 169-219, 2010-08-05.

Woods and King describe two characteristics of the Department of Defense intelligence communities post-World War II.

First, military intelligence (and counterintelligence generally) was not well defined and neither was its mission nor its operations. The National Security Act of 1947 effected some basic organizational structure for intelligence while assuming counterintelligence was included without specifically calling it out. The key terms *intelligence* and *counterintelligence* were not actually defined within the Act.

The second characteristic of military intelligence and counterintelligence was its full integration into the military mission of the Department of Defense. Counterintelligence operations were closely tied to actual military operations and were part of the standard military chain of command within the respective services: Army, Navy, and Air Force.

The body of the article is true to the title, assessing the oversight and control of DoD counterintelligence activities with specifics on declassified activities noted to support the authors' conclusions.

For example, the article describes the current (2010) state of DoD counterintelligence as one of "dystopic evolution." Counterintelligence has new elements of organization but not the capacity to co-ordinate its operations in a "symmetrical" manner. Force protection, homeland defense, and information fusion mission elements were fully assimilated into the operational culture of DoD counterintelligence while the balancing of these elements with protection of civil liberties within a legal culture remained a "pastiche of unintegrated authorities." Because of this, DoD counterintelligence operators have been able to choose their own priorities in the balance between service, mission, and protection of civil liberties. This might not be the most publicly desirable situation, especially if their service-biased choices are allowed to become ingrained in their operations.

These issues might directly account for the continuation of the counterintelligence campaign related to the Roswell Incident based on 1947 values, state of knowledge, and self-image of the Air Force. DoD counterintelligence components now are interacting more openly and directly with domestic civil society. There can be real concern for the present legal framework under which counterintelligence activities are conducted, and perhaps it is time for a comprehensive upgrade of that framework.

"Dystopic evolution." "Intelligence and counterintelligence undefined." "A pastiche of unintegrated authorities." "Choose their own priorities." Do these vague (and frightening) phrases help us understand how it might be possible for the US Air Force to continue its counterintelligence campaign related to the Roswell Incident on its own, without any accountability?

The Woods and King article addresses US Air Force counterintelligence (King is a former US Air Force Intelligence Officer) and provides a basis to understand the current situation. It may indicate the ability of the Air Force CI campaign to continue without public accountability or disclosure to protect the information of 1947. This overview also allows one to understand the potential of an ongoing CI campaign that restricts access to the Roswell Incident information within the US government, the Department of Defense, and even restricted within the US Air Force itself.

The need to know, a requirement for access to the information at any classification level, could be historically so highly restricted that it does not even include the President of the United States. While this may represent the "dystopic" nature of current CI affairs, it can be understood from a historical standpoint. In 1947 at the beginning of the UFO phenomena, the Air Force clearly struggled with how to handle information related to flying saucers, foo fighters, UFOs, and other potential threats to US airspace. It took a while for the Air Force to develop a policy and subsequent procedures.

Most likely, Colonel Blanchard's inadvertent disclosure came before the US Air Force decided to begin protecting information about UFOs, and it very well may have been the catalytic event that forced the Air Force into action, including the decades-long, ongoing counterintelligence campaign associated with the Roswell Incident.

Of course, it may also be possible that Colonel Blanchard, Major Marcel, Captain Cavitt, and others were just unable to properly identify that weather balloon, which was later "revealed" to be the top-secret Project Mogul data-gathering balloon. Spoiler alert, if you believe that, stop reading now.

I have two hypotheses about Colonel Blanchard's 1947 press release reporting the capture of a flying saucer.

Hypothesis #1: Colonel Blanchard was frustrated that the senior leadership of the Army Air Forces had not decided on policy or procedures regarding how to handle information concerning foo fighters, flying saucers, and UFOs. As a result, they had no policies, procedures, or guidance for him in July 1947. He found something highly unusual in the New Mexico desert near Roswell and apparently believed it was the wreckage of a "flying saucer." He probably sought guidance about how to handle this information, perhaps got tired of waiting for orders, and ultimately decided to make his own, command decision. That's what commanders do – they make decisions. This reasoning helps us to understand how he continued to have an exemplary career, rising to the rank of a four-star general and almost becoming the top leader in the United States Air Force. He exhibited leadership, exercised command, and most importantly, he kept the secrets.

Hypothesis #2: Colonel Blanchard had a wicked sense of humor and was just having some fun. Evidence from the US Air Force archives at Maxwell Air Force Base in Montgomery, Alabama, is the best place to go to find out about Colonel Blanchard's sense of humor. Sound recordings of Colonel Blanchard indicate a serious individual with a limited expression of humor. The most humorous recordings I was able to find were at Joint War Games, where he used humor to spar with his service rivals. Being an Air Force man, he deftly and on occasion humorously promoted the idea that only his service had the ultimate ability to defend the US in nuclear war and take nuclear war to the enemy. On the whole, however, the sound recordings I heard presented a military officer with a serious, straightforward composure who presented a limited expression of humor.

My conclusion? It is challenging to believe that Colonel Blanchard perpetrated a practical joke on the US Air Forces, the nation, and the world by issuing an imaginative and false press release announcing the recovery of "flying saucer" wreckage. If we dismiss Hypothesis #2, then that leaves us with Hypothesis #1.

CI 101: A COUNTERINTELLIGENCE PRIMER

Intelligence and counterintelligence are critical activities of the United States government and the US Air Force. Generally, in the broadest sense, and for the purposes of understanding the Roswell Incident, intelligence is the collection and use of information while counterintelligence is the protection of information, especially against other countries' intelligence efforts. While there is a plethora of information on the Internet about intelligence and counterintelligence, the *Introduction to US Counterintelligence: CI 101 – A Primer* document, written by Mark L. Reagan, Colonel, US Army (Retired) is an online, public document that is used as a reference for the rest of this chapter.

From the very beginning, the United States Army Air Forces and, later, the United States Air Force have been the US government organizations assuming primary responsibility for information about the

Roswell Incident. Even when opportunities for oversight of the US Air Force by other US government agencies were opened, the Air Force "volunteered" to lead the investigation efforts and managed to do so. The most glaring example was in 1994 when New Mexico Congressman Steven Schiff asked the Government Accounting Office to follow up on Roswell but ended up with US Air Force reports.

And, from the very beginning, the Air Force has chosen intelligence and counterintelligence personnel for leadership roles, operations, and control of the information about the Roswell Incident. This cannot be accidental. The first two US Army Air Forces personnel directly identified with the Incident are Major Jesse Marcel, 509th Bomb Group Intelligence Officer and Captain Sheridan Cavitt, resident (not 509th Bomb Group) Counterintelligence Officer.

When the US government finally, officially addressed the Roswell Incident directly in its 1994-1997 reports, it was Air Force intelligence/counterintelligence guru Colonel Richard Weaver in charge, not the Government Accounting Office, the US Congress, the Federal Bureau of Investigation, or the Department of Defense.

The language in the intelligence and counterintelligence fields differs from that used by astrobiologists, searchers for extraterrestrial intelligence, and so-called ufologists. This is important to see and understand the ongoing CI campaign. In fact, changing the language is one of the clues to discerning the counterintelligence campaign surrounding the Roswell Incident. Words and phrases such as *coverup* and *conspiracy* contribute to the success of the CI campaign to camouflage what happened.

As you read this book, be aware of how the words here are meant to reflect on intelligence and counterintelligence, not coverup or conspiracy. Language from the CI field describing CI activities includes, but is not limited to: *hinder, frustrate, exploit, negate, confuse, deceive, subvert, monitor, identify, neutralize, deter, manipulate, disrupt, destroy, military deception (MILDEC), control the clandestine collection operations, inadvertent disclosure, CI support to research technology protection (RTP), access agents, double agents, penetration operations.*

Are any of these words US Air Force descriptors? Can one read the US Air Force 1994-1997 Roswell reports without allowing for some of these CI activities?

A fellow named Michael Warner wrote an article for the CIA back in 2005 – "The Collapse of

Intelligence Support for Air Power, 1944-52: Two Steps Backward," *Studies in Intelligence*, Vol. 49, No.3, (2005). This article was available online at the Central Intelligence Agency official website.

The article caught my attention because of the title, including the dates covered. It helped me understand why the USAF can continue its sole service counterintelligence campaign concerning Roswell. It also verified my broad assessment of the priority of the three mission areas of the USAAF/USAF in 1947. The National Security Act of 1947 was the culprit.

In late 1945 President Harry S. Truman was probably looking to reform the intelligence communities. One can certainly understand his goal since he was suddenly thrust into the Office of President and having to make the atomic bomb decision without having been privy to the details of the Manhattan Project while serving as Vice President. One can also guess that, at the time, he probably had little background to understand the subtleties of what was being done in the name of intelligence and counterintelligence throughout World War II.

As he examined proposals for a new Director of Central Intelligence (DCI) to guide and coordinate the nation's intelligence activities, he agreed with the Army and Navy positions that every department needed control of its own intelligence. When he signed the DCI order in January 1946, it seems that he "punted" to the upcoming 1947 National Security Act. This allowed the services to continue their historic departmental intelligence practices, effectively allowing the independence of all the services to maintain and operate their own intelligence, even the new United States Air Force formed in September 1947.

This article helped me understand how the US Air Force was able to initially control the narrative around the Roswell Incident and begin the CI campaign that continues to this day. It also provides an understanding of how the USAF was able to control the 1994-97 efforts, the Condon Report, and even the most recent Department of Defense establishment of UAP organizations, missions, and activities. The Air Force's silence on UAP in the past ten years is deafening, isn't it?

OFFICE OF SPECIAL INVESTIGATIONS

Counterintelligence in the United States Air Force is generally handled by the Office of Special Investigations (OSI). When the US Air Force became a separate service in September 1947, the OSI did not exist. Counterintelligence was just counterintelligence. It wasn't until August 1, 1948, that the OSI was established, according to the organization's website document *2020 OSI Fact Book*. Most of the 2020 document talks about criminal investigations, and counterintelligence, appropriately, is very infrequently mentioned in the document. The public reporting about OSI proudly indicates that it was modeled after the Federal Bureau of Investigation (FBI) and that the first director in 1948 was recruited from the FBI. The final line in the *2020 OSI Fact Book* is so compelling to this book it has to be quoted here: "Find the truth." Challenge accepted. In March 2022, you could easily find this information online.

Page eight of the 2020 OSI Fact Book graphically reports the organizational elements of the OSI and includes "Offensive Counterintelligence Operations Division: Strategic Engagement; CI Referent; Community HUMINT LNO; SAF/AAH; Operational Source Validation; Psychologist." Lots to unpackage here. Or not.

"Offensive Counterintelligence Operations." Is this the organizational element that would handle a counterintelligence campaign for the Roswell Incident? Or is the information about Roswell so compartmentalized and so sensitive that it will never appear in writing or be assigned to a publicly named organization?

Maybe it really was just a weather balloon. Sorry, make that top-secret Project Mogul data-gathering balloon.

If the US Air Force Office of Special Investigations has been the organizational element responsible for the counterintelligence campaign, it has done an outstanding job since 1947, and just maybe there should be some positive recognition for this organization. To the best of my knowledge, this may be the longest running and most successful counterintelligence operation in history – except perhaps for those we knew absolutely nothing about until they were eventually exposed such as the Cold War's intrigues and deceptions revealed in the book, *Wilderness of Mirrors* by David C. Martin.

What crashed in the high desert near Roswell, New Mexico in July 1947? If it really was Project Mogul – and not an unidentified airship or spacecraft – none of the following deliberations in this section of this book really have any relevance whatsoever. However, if Colonel Blanchard and his folks correctly identified what they found and it represented a vulnerability in the defense of US airspace, then this section is important to understand the who, what, why, when, where, and how the vulnerability information might have been protected in 1947 and is still being protected. The key, of course, is counterintelligence.

"Neither Confirm nor Deny"

Neither confirm nor deny or NCND is a telltale sign that a counterintelligence effort is underway. This phrase originated with Project Azorian and the US Central Intelligence Agency (CIA) in 1974. A Soviet submarine went down in the Pacific, and the CIA put together an effort to recover the boat with the help of eccentric billionaire Howard Hughes. Hard to believe this stuff isn't made up. Hughes had a special ship built for the unprecedented deep-water recovery of the Soviet sub. She was named the *Hughes Glomar Explorer*, and the CIA first used the NCND phrase as part of the counterintelligence efforts to protect project information. This was the origination of the phrase, and it is sometimes referred to as "the Glomar response."

Initial use of the quote oftentimes represents a change in the CI campaign strategy and tactics. It could mean there really is something to the information (it's true), or it could mean that the campaign wants it thought that the information could possibly be true in order to initiate a response, even a nonresponse.

Does that last sentence confuse you? Sometimes CI efforts include release of classified information to confuse adversaries into thinking friendly capabilities are greater than they are. Sometimes the purpose is to distract adversaries into different thinking from the correct line of thought they were pursuing. There. That clears things up? Probably not, but this is an example of the deep, sometimes convoluted thinking required to understand a little of the CI world.

Presidential use of the phrase in connection to the Roswell Incident was specific when President Barack Obama used the phrase with Jaden Smith and his dad, actor Will Smith, while they were on a personal tour of the White House. It has been reported that Jaden asked the Roswell question of the President against the wishes of his dad. Of course, President Obama has a real sense of humor, and he may have just been joking when he used the NCND/Glomar response. Then again ...

On May 17, 2021, on *The Late Late Show with James Corden*, President Obama is the guest and is asked about UFOs. That wry wit of President Obama is on fine display when his answer includes: "When it comes to aliens, there's some things I just can't tell you on air." Sounds like a variation of the Glomar response? Whether you like his politics or not, there can be little argument about his comedic talent. You can watch the segment on YouTube.

When President Obama used the quote was it just an offhand remark intended to elicit a laugh?

Did it represent a change in the CI campaign strategy of overall denial? Was it just that wry, dry wit of President Obama? Does he really know the truth about Roswell? Did he have a need to know, as determined by the US Air Force?

The phrase means what it says. Denial or confirmation are both possible and sometimes both are used. The user may want to mislead toward denial of something that is true and needs to be protected. The opposite is also possible, the user may want to mislead toward disinformation being confirmed. No wonder former CIA counterintelligence chief James Jesus Angleton referred to counterintelligence as a "wilderness of mirrors."

OMISSIONS

A primary clue in seeing the continuing CI campaign is the omission of Roswell by the US Air Force in almost everything it publishes. In the entirety of the US Air Force's investigative efforts and public communications concerning UFOs subsequent to 1947, the Roswell Incident is almost always omitted.

Project Sign, Project Grudge, and Project Blue Book are the best publicly known Air Force UFO research efforts, and they all begin after the 1947 Roswell Incident. Yet none of them include Roswell in their findings. Unit histories, general officer biographies, and other public, official informational offerings from the USAF are consistent in their omissions of Roswell. Sometimes this even includes the omission of seemingly unnecessary offerings such as in the biography of Lieutenant General Roger Ramey.

In the 1960s, the US Air Force contracted with the University of Colorado to conduct a scientific, impartial, and external analysis of the Air Force's study of UFOs. At least that's what they originally claimed. The effort was rife with external criticism by academics who decried the selection of easily explained, readily apparent misidentification UFO sightings to the exclusion of difficult-to-explain sightings in its report. Of course, the Roswell Incident is not included. The history of the report contracting, the supervision of the work, and the publication of the final report do not include any exemplary examples of how to do research, analyze results, and come to conclusions without biasing the outcome from the beginning.

In its 1994-97 reporting on Roswell the US Air Force even brags on its success at removing Roswell from its UFO history. Here is an excerpt from *The Roswell Report: Fact versus Fiction in the New Mexico Desert*: "General Ramey's press conference and rancher Brazel's statement effectively ended this as a UFO-related matter until 1978, although some UFO researchers argue that there were several obtuse references to it in 1950s-era literature. Roswell, for example, is not referred to in the official USAF investigation of UFOs reported in Project Blue Book or its predecessors, Project Sign and Project Grudge, which ran from 1948-1969 (which Congressman Schiff subsequently learned when he made his original inquiry)."

While Projects Blue Book, Sign, and Grudge are cited, the University of Colorado's contracted report is not mentioned in the USAF's 1994-97 reports – a double omission?

The first chapter of this book describes the dedication of the Walker Air Force Base (formerly Roswell Army Air Field) civilian museum. While this was not an official US Air Force event, it included Air Force general officers, Air Force veterans, and veterans of the US Army Air Forces in addition to local civilians and other military personnel. The omission of the Roswell Incident during the event, including omission of any note whatsoever at the new museum itself, speaks volumes to the success of the US Air Force in its counterintelligence campaign to omit the Roswell Incident from history.

In addition, the official US Air Force biography of General Blanchard (Colonel Blanchard in 1947) omits any reference to the Roswell Incident. And, notably, the current official Air Force unit history of the 509th Bomb Group omits any reference to the Roswell Incident.

From the US Air Force's perspective, why would it reference an Incident that was the result of a mistaken identity of a weather balloon? In the history of Walker Air Force Base (formerly Roswell Army Air Field) were there just so many more significant historical events that the July 1947 Incident was not important? Remember, Roswell Army Air Field was tasked with operational deployment of nuclear weapons – do nuclear weapons and UFOs represent an unthinkable combination? Will the continuing omission of Roswell in US Air Force investigations and reports eventually result in omitting the Roswell Incident from history and from our collective memory?

ABOVE ALL

Like all the US military services the United States Air Force has a slogan. Over the years the slogan has changed, sometimes reflecting the times and sometimes for apparently no reason other than change itself. Perhaps it is hoping to find one that resonates. Here are some of the US Air Force slogans:

- A Great Way of Life
- Above All
- We've Been Waiting For You
- It's not Science Fiction, It's What We Do Every Day
- Cross Into The Blue
- No One Comes Close
- We Do the Impossible Every Day
- Do Something Amazing
- The Sky's No Limit
- Fly-Fight-Win
- Ano Ab Alto (One From On High)
- Aim High
- Death From Above
- Defenders of the Force
- Excellence in All We Do
- Follow Us
- Guardians of the North
- Integrity First
- Liberty We Defend
- One Team. One Mission.
- Peace Is Our Profession
- Return With Honor
- Service Before Self
- These Things We Do, that Others May Live

My favorite is Above All. Succinct, strong, and arrogant. It didn't last long, sort of like the post-WWII cooperative, interservice, combined intelligence efforts and organizations.

These slogans should give you an impression of the US Air Force: how it sees itself, what it thinks it is all about. There are several sources online that document these slogans, but you may have to visit more than one to compile your own list like I did.

SEEKING TRUTH

Visiting Roswell to seek truth about the seventy-five-year-old Incident will probably not satisfy most inquisitors. When you visit the International UFO Museum and Research Center on Main Street and attend the annual celebration held in July, you will find that the preponderance of information is clearly on the Believer side of the continuum. There were so many "witnesses," non-Roswell stories, and other distractions that the official US government side of the story seems to be the only place to seek out the truth. However, the omissions, contradictions, and changes in the stories over the years by the Air Force prevent the rational, deductive-reasoning seeker from attaining resolution.

After living in Roswell for eight years, taking two separate one-week visits to the National Archives in Maryland, spending a week at the US Air Force Historical Records at Maxwell Air Force Base in Alabama, driving to Colorado for a couple of visits to the US Air Force Academy Cemetery in Colorado Springs, and taking numerous trips to the high desert northwest of Roswell to visit the "crash site," it seems readily apparent to your author that the US Air Force has been conducting a CI campaign from July 1947 to the present day to protect the information about the Roswell Incident. Air Force counter-intelligence officials have been successful for seventy-five years. We probably will not know the truth until they decide to divulge it.

Personal Disclosure

Your author is not an intelligence or counterintelligence professional and has never worked in either field, but I did spend almost thirty years in the Department of Defense, most of it with security clearances to work on classified projects.

Two career experiences stand out that might influence this writing. First, I was on a Special Access Project with the Air Force where I identified system vulnerabilities in a classified memo that documented my work. The US Air Force classified the memo such that I was not allowed to read it post-authorship. This memo identifying *vulnerabilities* was compartmentalized (classified) at a level above my need to know. In other words, because the memo discussed vulnerabilities it was classified at a high level, and I was not allowed to read the memo I wrote.

One other career experience is notable here. During the height of the Cold War, I worked on a project in Europe that required a special NATO communications clearance. My in and out briefings with the project very definitively informed me of the sensitivity of one particular class of information – information associated with *vulnerabilities*.

Based on these two experiences, it is easy for me to understand the US Air Force's continued efforts to protect information about any vulnerabilities that may have been identified in 1947 and continue today. The same is true of any information regarding frequencies, types of energy, new engineering concepts, or even basic information either confirmed or refuted by events of the Roswell Incident.

In the end, what I see is a counterintelligence campaign that began on July 9, 1947 and continues to this day.

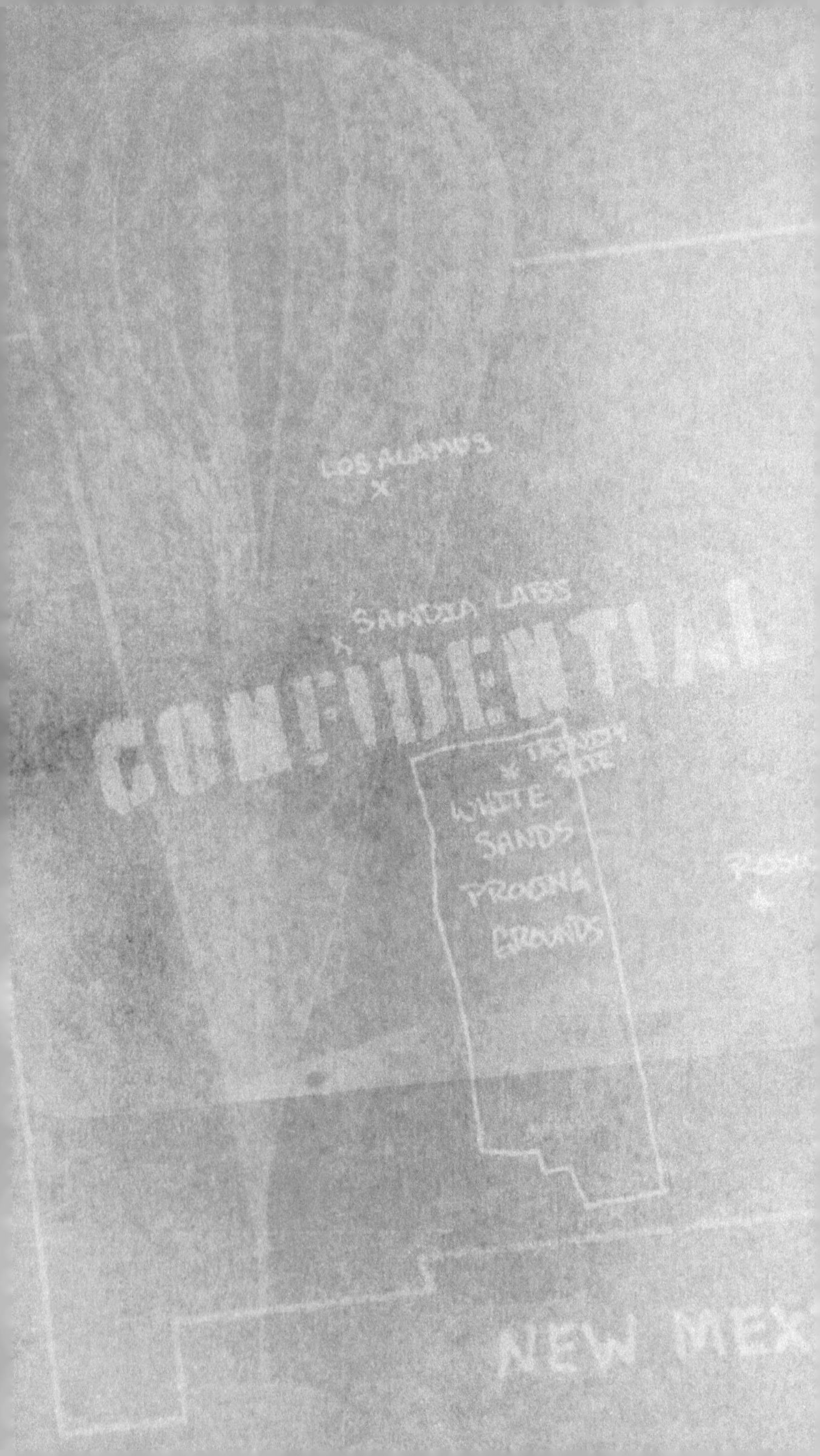

LOS ALAMOS
X
X SANDIA LABS
CONFIDENTIAL
Y TRINITY SITE
WHITE
SANDS
PROVING
GROUNDS
NEW MEX

CHAPTER 5
The Search for Extraterrestrial Intelligence

AMERICAN ASTRONOMER AND ASTROPHYSICIST Carl Sagan repeatedly noted that even if we don't have tangible evidence of extraterrestrial life this doesn't mean that it doesn't exist.

Your author struggled with the decision to include this chapter. In the end, it had to be included because to hypothesize that the US Air Force continues to protect information and evidence gathered in 1947 requires one to consider exactly why it is necessary, seventy-five years later, to continue the CI campaign. The most logical and simplest explanation is that evidence of extraterrestrial life with capabilities far beyond the capabilities of the US Air Force to counter must be protected, because this currently represents a vulnerability of the United States and the Air Force.

Naturally, this vulnerability hypothesis requires acceptance of the hypothesis that intelligent extraterrestrial life exists. So, here we go with a brief argument for accepting the hypothesis that the existence of extraterrestrial intelligence may be true.

The consideration of extraterrestrial life has been a part of the human experience for most of recorded history, and most of the conflict in accepting the validity of this hypothesis has been with religion. For instance, the Catholic church's early doctrine and belief structure didn't provide for wide variances in the cosmology of the universe being different from its theological views. Probably the best-known example is that of Galileo Galilei, a seventeenth-century Italian astronomer and physicist. Based on his celestial observations, he championed something called *Copernican heliocentrism* – the idea that the Earth revolved around the sun and, shockingly, the Earth was not the center of our solar system or the universe. The Catholic church tried and punished Galileo for his heretical ideas, which contradicted Holy Scripture. There are numerous other examples, but they are not directly applicable to the events of July 1947.

In 1947 there was no National Aeronautics and Space Administration (NASA), no Very Large Array

Radio Telescope in New Mexico, no *Voyager 1* spacecraft leaving the universe with our best guess of a message interpretable by extraterrestrial intelligence, no Arecibo interstellar radio message to be sent outward, no Mars Exploration Rovers, and so forth. The public knew little about radar, nuclear weapons, and the like. Those were leftover secrets from the recently won war. ET was a subject for science fiction, emphasis on *fiction*.

Carl Sagan and American astronomer and astrophysicist Frank Drake were some of the early pioneers in the search for extraterrestrial intelligence, also known as SETI. Drake even came up with an equation that is still used today to judge the probabilities of extraterrestrials: the Drake Equation. These efforts began in the late 1950s and early '60s. There are plenty of online, library, and video sources if you are interested. If nothing else, you may want to see the current probabilities associated with the Drake Equation.

The history of NASA's SETI efforts seems to be two steps forward and one step back, starting in 1958 with the establishment of our historic space organization during the Eisenhower administration. NASA had several programs, some working with outside SETI adventurers, but finally in October 1992, nearly thirty-five years after its establishment, NASA initiated an actual project, called SETI. There were lots of publicly noted reasons for Congress cancelling the fledging program the very next year, but the public may never know the true reasons.

Today you have only to watch the news on television, search online, or follow NASA live streams to keep up with the many NASA programs looking for extraterrestrial intelligence.

The search for extraterrestrial intelligence, or SETI, has moved from science fiction to true science in the seventy-five years since the Roswell Incident in 1947. Today there is an almost weekly discovery that brings us humans closer to discovering that there is life elsewhere in the universe with a good probability that some of it is intelligent. Money (lots of money) is being spent on astronomy, planetary exploration, astrobiology, and astrophysics in record amounts by NASA and other space agencies around the world.

Literally thousands of potentially habitable planets have been discovered. In February 2017, a press release from NASA reported the discovery of seven potentially habitable planets astronomically close to Earth with three of the seven potentially having that essential of life as we know it: water.

Perhaps the disclosure of what was really found in the New Mexico desert in 1947 by the Air Force will be forthcoming too?

One of the best discussions about the possibility or probability of extraterrestrial life and UFO/UAP is from a field you may not have thought about – risk management – as in disasters, natural emergencies, and threats to humanity.

In an article that appeared in the May 2014 journal *Risk Management*, "Preparing for Extraterrestrial Contact," author Mark Neal begins by pointing out one of the strategies of the CI campaign – the marginalization of most academic and/or scientific research except in the fields of astrobiology, astronomy, and sociocultural phenomena. While there is ever-increasing scientific evidence that supports the idea of the possibility of extraterrestrial intelligence, serious academic and scientific efforts including risk management study still are denigrated, degraded, and discounted. He also points out that one of the areas especially subject to defamation and belittlement is politics.

Using a risk management approach, Neal postulates five scenarios upon which to base disaster prevention and management efforts:

Scenario #1: There is no extraterrestrial intelligence. We are the only "intelligent" life.

Scenario #2: There is extraterrestrial life, but it is not intelligent.

Scenario #3: There is intelligent life on other planets, but it is technologically inferior to us.

Scenario #4: There is intelligent life on other planets, which is technologically superior to us.

Regarding this fourth scenario, Neal says, "If such extraterrestrials are benign, then they might contact us beforehand. If their intentions are mercenary, however, they are likely to remain silent, so as not to give us time to prepare for the encounter. The silence at the heart of the Fermi Paradox may reflect stealth, not absence." (There are plenty of sources online to explain the Fermi Paradox, but few quote Enrico Fermi directly. It might be because he made the remark over a lunch break with colleagues, and his words have been passed along through word-of-mouth. Here is your author's take: It is highly probable that given the age and size of the universe, there is and has been other intelligent life. If so, where are they?)

Neal continues: "Ambiguity and uncertainty about what to do, or how to respond, may thus characterize a first encounter. This lack of agreed protocol itself poses risks for humanity.

... At present, there is a strategic insularity in military thinking and preparedness, which sustains a vulnerability to external third-party threat."

Scenario #5: They have already been here/they are already here.

Neal concludes, "We have reached a stage in scientific research whereby exoplanets able to bear life have been identified, and major scientific programmes are seeking evidence of extraterrestrial life itself. At the same time, however, the prospects of contact, visitation, or indeed the presence of aliens are still treated as peripheral, marginal issues by government, policymakers, and social scientists."

Can we add that the United States Air Force may be at the forefront of marginalizing the possibility or probability of intelligent extraterrestrial life, and thereby not allowing us to prepare to deal with the potential disaster of a close encounter? Has the USAF already moved on to risk management and disaster prevention from a defense, counter-force perspective and just hasn't shared its efforts?

I miss comedian George Carlin. I especially miss his ability to look at things differently – to see things that are hidden in plain sight but not obvious to the rest of us. His comparison of belief in UFOs (aliens) and Jesus is important and relevant to addressing Roswell as well as being entertaining.

Carlin directly compares the two beliefs and includes the public, general acceptance of one over the other. It isn't rocket science to figure out which one is generally accepted, and which one is generally discounted. Carlin makes fun of both, and at the end of this monologue he picks a "winner." His contrast of the two beliefs is focused on their similarities and the absurdity of honoring one while disclaiming the other.

He points out that one is based on unproven, undocumented mythological non-data. Can you guess which one?

In addition to stimulating your cognitive processes while being entertained by one of America's greatest comedians, listening to his monologue would be a great break from being too serious about this topic. You can find it online in its entirety, and it is highly recommended. The video recording is much preferable, but the written transcript is better for detail. Do an Internet search for "They Came from Out of the Sky," by George Carlin.

In memoriam of George Carlin and inspired by his insights, the following quiz is offered.

What is the difference between extraterrestrial believers and Jesus followers?

- There is no difference.
- One is based on science.
- The US Air Force hasn't lied about one.
- The US Air Force requires a security clearance to talk about one.

SUMMARY

1. Our understanding of the possibilities and probabilities of extraterrestrial life, including intelligent alien life, continues to increase as science seeks it out.

2. There is currently no generally known physical evidence of extraterrestrial life.

3. If we don't search, the probability of finding evidence is 0.0.

4. Absence of evidence is not evidence of absence.

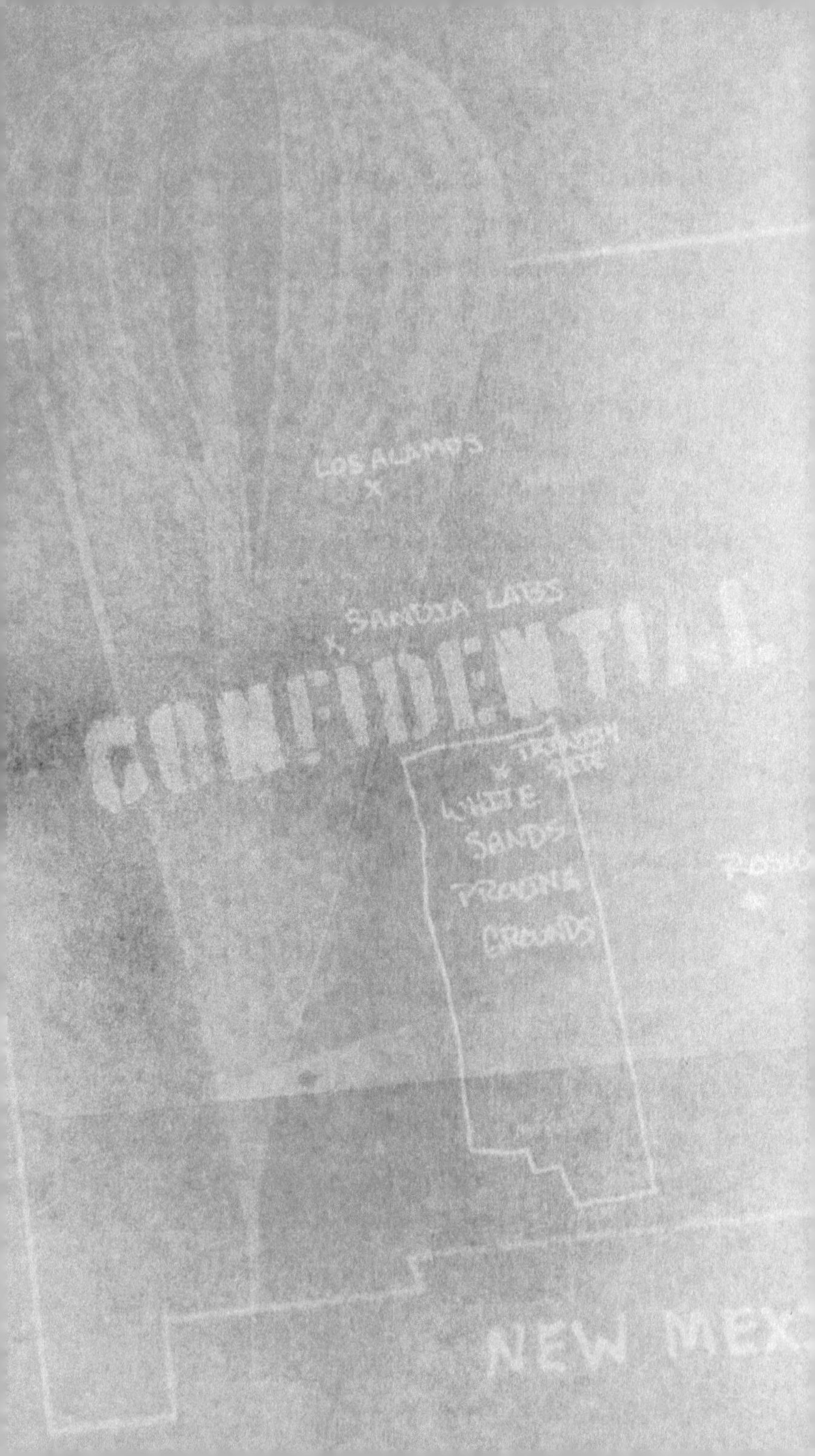

LOS ALAMOS
X
SANDIA LABS
X
CONFIDENTIAL
X TRINITY SITE
WHITE
SANDS
PROVING
GROUNDS
NEW MEX

CHAPTER 6
Colonel William H. Blanchard

WILLIAM HUGH BLANCHARD was born in Chelsea, Massachusetts, in March 1916. The 1930 US Census shows him in a household that included his mother and his grandfather. Blanchard graduated from prestigious Phillips-Exeter Academy and went on to attend the United States Military Academy West Point ... blah, blah, blah.

Someone should write a decent biography of this man! His official USAF biography online does not do justice to his career or give a complete picture of the man. Searching for documents and details of his life is a frustrating and incomplete exercise. Parts of his life story are scattered about from Boston to personal anecdotes in USAF unit histories as well as a limited collection of his personal papers at the US Air Force Historical Library, Maxwell Air Force

Base, Montgomery, Alabama. While these sources of data about his life may be adequate for a traditional biography, assessing his demeanor – including his sense of humor – is more difficult.

USAF General Blanchard biographical photograph

Colonel William H. "Butch" Blanchard was in command at Roswell Army Air Field in 1947, ordered the original press release that started the "hurricane" of Roswell, and should be the eye of the storm that continues to this day, despite the Air Force's best efforts to divert attention from him.

Blanchard's career in the US Army Air Forces/ US Air Force began with his acceptance to the US Military Academy at West Point, New York, in 1934.

He rose to the rank of General (four stars) and died of a massive heart attack while on active duty at the Pentagon as the US Air Force Vice Chief of Staff, an almost certain choice to be named the next Chief of Staff of the US Air Force had he lived.

Graduating from West Point just before World War II, Blanchard was initially a Field Artillery Officer before he became a pilot in the Army Air Corps, later the US Army Air Forces. He must have been a pretty good pilot since he remained at the USAAF pilot training bases around San Antonio, Texas, as an instructor after he got his wings. His flight records are in the archives at Maxwell Air Force Base, and they show he was an active, flying command pilot up until his death.

During WWII he was part of the B-29 bomber efforts almost from the beginning. The B-29 was the most advanced aircraft in the world in 1943, and Boeing did its best to produce them quickly and in great numbers in Kansas. The rush to produce the weapon system that included the mating up of trained air crews and their new aircraft was informally known as "The Battle of Kansas." Initially assigned to the training bases in Kansas where the B-29s came off the assembly lines and were turned over to combat crews, Blanchard later flew the bombers to India, China, and the Pacific theater where he rose in rank to Colonel.

He flew combat missions over Japan and worked for General Curtis LeMay, planning the operational missions for the two atomic bomb strikes at Hiroshima and Nagasaki. The end of the war found him uniquely positioned to command the 509th Composite Group (later known as the 509th Bomb Wing at Roswell Army Air Field), succeeding Colonel Paul Tibbets, the first 509th commander who piloted the Hiroshima Atomic Bomb mission.

The 509th was the world's first atomic weapon capable military unit, and it found itself in Roswell, New Mexico, after the war. Colonel Blanchard certainly had security clearances appropriate to the tasks of commanding the world's only nuclear capable unit, planning atomic bombing missions, and building up the fledgling Strategic Air Command of the Air Force.

His post-1947 assignments contributed to his rise in the Air Force and seem to have been a "flight path" to the highest position in the US Air Force. As Inspector General of the Air Force in the 1970s, he had the unenviable, politically sensitive duty of handling the famous TFX Fighter program promoted by then Secretary of Defense Robert McNamara. In addition to the political intrigues of the program, there were additional complications associated with interservice rivalries between the US Air Force and the US Navy.

The success of his career cannot be questioned whether before or after the 1947 Roswell Incident.

There is a nagging question concerning Blanchard: Why is there a unique, standalone memorial for General Blanchard at the US Air Force Academy Cemetery? While his career story is notable, interesting, and historic, it is his final resting place that offers present-day mystery alongside the older mysteries of Roswell.

General Butch Blanchard is interred at the United States Air Force Academy Cemetery in Colorado Springs, Colorado, along with some of the USAF's most notable leaders. General Curtis LeMay, who led the "Mighty 8th" Air Force in Europe during the second World War before taking command of the B-29s that pulverized Japan and dropped the atomic bombs at Hiroshima and Nagasaki is there, right next to Blanchard. US Army Master Sergeant William J. Crawford, WWII Medal of Honor recipient, Combat Infantry Badge holder, and former janitor at the US Air Force Academy is there as well – he is the only US Army enlisted soldier in the cemetery and the only Medal of Honor recipient. Not too far from

Blanchard is General Carl "Tooey" Spaatz, the very first Chief of Staff of the brand-new US Air Force in 1947 and former WWII European strategic bomber as well as WWI aviator. US Air Force Academy Graduate Second Lieutenant Laura Piper was killed in Iraq and is buried at the Academy, where she was also born.

The US Air Force Academy Cemetery is a tidy, well-organized place, and there are restrictions on stones, markers, headstones, plaques, cenotaphs, and memorials. No matter the rank, gender, or history of the decedent, all the memorials and markers are alike in design. Only one edifice stands apart from the uniform, shades of gray, approved markers and memorials in the cemetery.

General Blanchard's memorial at the
US Air Force Academy Cemetery.

Why was Blanchard allowed the special, one-of-a-kind memorial?

General Blanchard's grave marker at the
US Air Force Academy Cemetery.

In November 2016, the ground around the special Blanchard Memorial had been resodded as if some recent digging had been done. When I queried a friend who is a retired Air Force combat pilot and US Air Force Academy graduate about the Blanchard Memorial, he immediately retorted, "Where do you think they hid the evidence?"

On a visit to the Cemetery, I met with the United States Air Force Academy Mortuary Officer at the Cemetery. When asked about the Memorial for Blanchard, she reported that she did not know why it was the only different marker. She promised to look through her files including the documents directly related to the interment service for General Blanchard in 1966. She promised to call or email me with any information she found. She also reported that the disturbed grass around the marker was due to landscaping activities to upgrade the area since it was in a shady spot and did not get enough sun to maintain the grass. When I had not heard from her after thirty days, I called at least three times, leaving voicemails, and I emailed her at least three times without any response. Finally, I sent two letters – without response.

When I first penned this section of the book, I began to wonder if there was anything of substance to my retired Air Force pilot friend's comment. Could the Roswell evidence really be buried within General Blanchard's unusually large memorial? My first approach was sarcastic humor, borderline encouragement of this conspiracy theory. Later, I considered deleting the section entirely. Finally, I realized that it had to be included because this is an excellent example of what I perceive as causal for two successive actions:

1. Conspiracy theorists highlight the anomaly, and here we go down another rabbit hole of speculation.

2. The US Air Force decides not to respond or explain (until a congressman inquires?), creating a vacuum that allows for another distraction of incredulity, fancifulness, and deviation from pursuing the original questions associated with "flying saucer versus balloon." What I see here is the potential to fall into the trap of speculation leading to more craziness. No thank you.

General Blanchard is a key figure in the Roswell Incident, and he was a decorated, brave, WWII leader who had pivotal roles in major efforts against the Japanese. His career spanned the gap between the traditional warfare of WWII and the mutual assured destruction military strategy of the Nuclear Age. He was a key figure in the Cold War and a public figure for the Department of Defense in the political arena.

Perhaps here is a good place to reiterate a thought from the beginning of this chapter: Someone needs to do the research and write a thorough, honest biography of this warrior and historical figure.

I don't know why General Blanchard's memorial is different from all the other structures – it is unique and imposing. I don't know why the US Air Force

Academy Mortuary Officer never communicated with me. And I will not assume any responsibility for promoting this conspiracy theory.

But what about the nature of Blanchard the man? Did he have a wicked sense of humor that might enable us to interpret the Roswell press release as some kind of joke? As I mentioned earlier in this book, I listened to multiple voice recordings of General Blanchard obtained from the US Air Force Historical Research Agency at Maxwell Air Force Base, Alabama. These offer some clues to his sense of humor and whether or not it may have contributed to reporting recovery of a flying saucer in 1947.

The assessment of his sense of humor is difficult. Why assess his sense of humor? As noted earlier in this book there are arguably two hypotheses that could possibly explain why Colonel Blanchard issued that now-famous Roswell press release in July 1947:

1. He was telling the truth. He may have sought guidance from his superiors about how to communicate the evidence that was found. Perhaps, after waiting for a period of time and receiving no guidance, he made a command decision to issue the press release. During my more than thirty years with the Department

of Defense, I witnessed this strategy on at least three occasions by general officers with decisions that would have notable consequences. The principle they used was: "It is easier to ask forgiveness than to get permission." Colonel Blanchard made a command decision. It is what commanders do. His forgiveness is evidenced by his subsequent successful career and the attainment of the rank of General. Certainly, the Roswell Incident spurred the Air Force into action, and this is one of the main reasons to identify ongoing Air Force efforts concerning Roswell as counterintelligence.

2. He had a wicked sense of humor, and the press release was a joke.

My definitive take on this question is no, Blanchard did not have a wicked sense of humor. As a serious military commander, he would not issue a "practical joke" press release. This conclusion was based primarily on a lack of evidence of a generous sense of humor in the public records associated with Blanchard and research specifically conducted at the US Air Force History Agency, Maxwell Air Force

Base, AL. During my visit in June 2015, the Agency provided me with recordings of a speech General Blanchard gave at the Air War College. These recordings revealed a sense of humor used primarily for banter in interservice conversations.

General Blanchard's official online US Air Force biography is notable for its omission of the Roswell Incident. While most of the world seems to recognize his name for the Roswell Incident, the US Air Force doesn't even mention it.

Blanchard's name comes up in the 1994-97 USAF Roswell reporting. Most of the citations appear to provide him with alibis for not being there (he went on leave on July 8, 1947, which was approved on July 8, 1947), the fact that he was not aware of Project Mogul (the data-gathering balloon), and later, as USAF Inspector General, for being a great guy.

Without descending into the world of eyewitnesses, hearsay, and pure speculation, it is difficult to reconstruct Colonel Blanchard's hourly activities, decision making, and movements during the initial period of the Roswell Incident between July 2 and July 8, 1947. Certainly, the specific time that he signed out on leave on Tuesday, July 8, 1947, can only be guessed. After the press release and the firestorm global response that centered on Colonel

Blanchard, Roswell Army Air Field, and the Army Air Forces following the *Roswell Daily Record* article and Associated Press story, can we really imagine that his actions that day didn't include multiple conversations with his superiors, subordinates, and a broadening sphere of "friends," mainly from the intelligence and counterintelligence communities?

Finally, did Colonel Blanchard's press release force the hand of senior US Air Force leadership to establish policy, procedures, and strategies to address the existence of UFOs and the potential threat to the United States by technically superior forces operating in our airspace? Does that possible vulnerability continue to this day?

LOS ALAMOS
X
X SANDIA LABS
CONFIDENTIAL
X TRINITY SITE
WHITE
SANDS
PROVING
GROUNDS
ROSW
NEW MEX

CHAPTER 7
Colonel Richard L. Weaver

OUTSIDE OF THE UFO COMMUNITY and the relatively closed fraternity of Air Force Intelligence/ Counterintelligence, Colonel Richard Weaver's name is not well known. In the history of the continuing Air Force counterintelligence campaign, Colonel Weaver is a key figure. Arguably, he is *the* key figure from 1994 until his retirement in 2001.

His personal account of his 1994-1997 official duties investigating and reporting on the Roswell Incident are detailed in his September 2020 book: *Backstory: Roswell: Exclusive Untold Disclosures about the 1994 Air Force Roswell Report Told by the Man Who Led the Inquiry.*

When Congressman Steven Schiff from the 1st Congressional District of New Mexico initiated a Government Accounting Office investigation of the

Roswell Incident in 1994, Colonel Weaver was selected by the US Air Force to head up the response and ended up writing the final report: *The Roswell Incident: Fact versus Fiction in the New Mexico Desert.* It was initially intended to be an independent, external investigation by the Government Accounting Office. How did the US Air Force end up investigating itself and writing the final report?

One of the strongest pieces of evidence for me in concluding that the Air Force is continuing to carry out a counterintelligence campaign to protect the Roswell information is the recording of Colonel Weaver's interview of Lieutenant Colonel (Retired) Sheridan Cavitt. The transcript is included in the 1995 *Fact versus Fiction* report.

A captain at the time of the Incident, Cavitt served as Counterintelligence Officer at Roswell Army Air Field in 1947. The interview was conducted by Colonel Weaver at Cavitt's home in 1994 in the presence of Cavitt's wife. The transcript of the interview is published in the report, and you may struggle to accept it as an attempt to reach the truth if you begin with an open mind. It is available to the public as an audio recording at the National Archives and Records Administration facility in College Park, Maryland (Archives II). I have listened to the recording numerous times as well as reading the transcript in the 1995 *Fact versus Fiction* report, both before and after visiting the NARA.

Listen to this interview, and you may come to the same conclusion that I did. It is an interview that is blatantly obvious in seeking a predetermined outcome. That predetermined outcome did not include verification of any flying saucer recoveries.

It is the kind of "evidence" that a skilled prosecutor would most likely rip to pieces in cross examination in a courtroom trial. A gentle reminder here is that Colonel Weaver was part of the Air Force Office of Special Investigations, modeled after the FBI and touted as a superior investigative service.

Colonel Weaver's career in the US Air Force is difficult to document, but this is not surprising for any counterintelligence officer. Here are highlights from what was reported in an online source in 2015 that is no longer available:

Colonel Weaver was on active duty with the US Air Force for twenty-eight years, most of it as a Special Agent with the Air Force Office of Special Investigations. He held command positions with four separate investigative field units. One of his last positions was Director, Security Policy and Special Program Oversight, Office of the Secretary of the

Air Force. He held this title when he conducted the Roswell Incident investigations of 1994-1997 and wrote the *Fact versus Fiction* report.

He was perhaps *the* recognized expert in the Air Force for counterintelligence efforts for military force protection efforts. He held both a bachelor's degree and master's degree in forensic sciences. He was inducted into the Air Force Office of Special Investigations Hall of Fame. His was a distinguished career with a multitude of successful assignments to highly visible and well-known efforts including leading the Khobar Towers bombing investigation in 1996 (a terrorist attack on a USAF housing complex in the town of Khobar, Saudi Arabia).

In retirement, Colonel Weaver had a security consulting business and did volunteer work, but he generally maintained a low public profile.

As noted, in 2020 he wrote his own book about his involvement in the Roswell Incident: *Backstory: Roswell: Exclusive Untold Disclosures About the 1994 Air Force Roswell Report Told By The Man Who Led The Inquiry.*

There are two key takeaways from his book for me. First is the classification of the two groups he identifies associated with Roswell: The "Sons of Light," identified with the US Air Force, and the

"Sons of Darkness," identified with New Mexico Congressman Schiff who initiated the investigation of 1994, UFO researchers, and the US Government Accounting Office.

My second key takeaway is the "beginning" of the story being held up as the weather balloon with the "revelation" that it was actually a top-secret, data-gathering balloon called Project Mogul. As noted in the early pages of this book, the flying saucer beginning has already seen omission.

There is little doubt Colonel Weaver would see the book you are reading as being in the "Sons of Darkness" camp. I will not return the favor with a derogatory label. By authoring both the *Fact versus Fiction* report and his *Backstory: Roswell* book, Colonel Weaver was doing what he thought was in the best interest of the US Air Force in continuing the counterintelligence campaign.

Weaver's personal book about Roswell was written after his retirement from the US Air Force. Did he need the money? Did the Air Force approve the project and support it to continue the counterintelligence campaign? Do you really want to spend $17.50 on Amazon to re-read what is essentially the 1995 *Fact versus Fiction* report? After reading the book, do you think you would put yourself in one of the two groups that Weaver identifies: the "Sons of Darkness" or the "Sons of Light"?

There are two voice recordings that played a key role in reaching my conclusion that the Air Force is still conducting a CI campaign concerning the Roswell Incident, and Colonel Weaver's interview of Lieutenant Colonel (Retired) Sheridan Cavitt is one of them. It is transcribed within the almost 1,000 pages of the 1995 report, but listening to the audio recording at the National Archives in College Park, Maryland was infinitely more rewarding than reading the written words.

Mrs. Cavitt's "help" in the interview, Weaver's obviously leading questions, and Cavitt's struggle with a fading memory are clearly discernable. By any reasonable interpretation, the interview was not an impartial effort to determine truth, but an investigative, inductive reasoning effort to support a desired, established, predetermined storyline. One only need remember that Colonel Weaver is an experienced investigator conducting the interview to understand his potential motives for leading his witness, structuring his questions by telling a story and asking if it was correct, and encouraging Mrs. Cavitt to "help" when Cavitt struggled with his answers.

The Air Force wants you to listen to this interview or read the transcript and conclude that the Roswell find was a balloon. I came to the opposite conclusion.

A visit to College Park, Maryland, should be favored over a visit to the International UFO Museum in Roswell, New Mexico, by those seeking the truth.

Incidentally, the other voice recording that played a key role in reaching my conclusion that the Air Force is conducting an ongoing counterintelligence campaign concerning the Roswell Incident is the recording of General Blanchard at a Pentagon War Game, which the US Air Force Historical Agency graciously provided me.

LOS ALAMOS
X
SANDIA LABS
X
CONFIDENTIAL
TRINITY SITE
WHITE SANDS PROVING GROUNDS
NEW MEX

CHAPTER 8
Colonel Jesse A. Marcel, Jr.

THE US AIR FORCE doesn't talk much about Colonel Jesse A. Marcel, Jr., MD. Maybe it's an interservice rivalry thing, since he was a US Navy Lieutenant and Physician as well as a US Army National Guard Colonel, Flight Surgeon, and Combat Pilot. Certainly, his inclusion in the 1995 *Fact versus Fiction* report is meant to discount him as a witness in the Roswell Incident.

Colonel Jesse Marcel's father was the intelligence officer – Major Jesse Marcel – who was originally sent to the crash site northwest of Roswell in July 1947. On his way back to Roswell Army Air Field from the site with some of the crash debris, Major Marcel stopped by his home in Roswell late at night where he woke up his wife and eleven-year-old son, Jesse. He shared the materials with them before he returned to Roswell Army Air Field. He and his son performed personal examinations of the various materials including trying to burn them and attempting to cut them – unsuccessfully.

Colonel Jesse Marcel, MD (son of the late Major Jesse Marcel and eleven years old at the time of the Incident) stated the following in an affidavit dated May 6, 1991: "There were three categories of debris: a thick, foil-like metallic gray substance; a brittle, brownish-black plastic- like material, like Bakelite; and there were fragments of what appeared to be I-beams. On the inner surface of the I-beam there appeared to be a type of writing. This writing was a purple-violet hue, and it had an embossed appearance. The figures were composed of curved, geometric shapes. It had no resemblance to Russian, Japanese, or any other foreign language. It resembled hieroglyphics, but it had no animal-like characters." (This excerpt from the affidavit was published in *The Roswell Report: Fact versus Fiction in the New Mexico Desert*.)

If you read just the above quote, you would have no trouble believing statements from counterintelligence Colonel Weaver about the nature of the balloon debris in the *Fact versus Fiction* report. Unfortunately for us, there are almost 1,000 pages with pictures of balloons, data charts for wind speeds in the Tularosa Basin, drawings of tape with shapes on them, and lots of affidavits saying no one had any records of flying saucers.

Colonel Weaver did not conduct an interview with Colonel Marcel for this report, and there is no transcript of the May 6, 1991, interview with him. Turns out the 1995 Fact versus Fiction report only used a selectively edited quote from another source, an author writing on the Believer side of the continuum. Wonder why Weaver didn't include "Colonel" with his identification of Montana State Surgeon General, Colonel (Retired) Doctor Marcel? More curiously, why didn't he interview Colonel Marcel himself for his 1,000-page, "comprehensive" *Fact versus Fiction* report?

Colonel Marcel, the son of Major Jesse Marcel, the 509th intelligence officer who originally went to the crash site, told the same story about the Roswell Incident from childhood throughout his entire adult life until his death in 2013. He was consistent, explicit, and unwavering in his storytelling. His story is one of personal, hands-on exposure to materials "not of this world."

"Not of this world" – these are Colonel Marcel's words, as quoted directly from his father. Do you wonder why these words weren't included in the quote used in Colonel Weaver's *Fact versus Fiction* report? Me too.

Colonel Marcel's military career included service as a physician in the US Navy during the Cuban Missile Crisis and later in Vietnam, Flight Surgeon in the Montana Army National Guard, US Army Helicopter Pilot, and a thirteen-month combat tour in Iraq with the US Army 189th Helicopter Battalion. Colonel Marcel flew 225 combat hours in Iraq. Somehow, he managed to get appointed as Montana State Surgeon General along the way.

My guess is that he had more "fruit salad" (military decorations, awards, and medals) on his dress uniform than Colonel Weaver ever had.

It is difficult to understand why Colonel Weaver and the US Air Force didn't include him in the report's investigative efforts, unless there is an ongoing CI campaign, and they couldn't figure out a way to discredit Colonel Marcel tactfully but realistically. Even the United States Air Force hesitates to attack the character of a commissioned officer directly and publicly, even if he was in a rival service.

In 2008, Colonel Marcel published his own book about his 1947 Roswell experiences: *The Roswell*

Legacy: The Untold Story of the First Military Officer at the 1947 Crash Site.

Throughout his life, he was specific, believable, and consistent in his story about the materials he handled with his parents in 1947. Just like his dad, he concluded the materials were not a weather balloon – they were "not of this world."

Colonel Marcel was never interviewed for the US Air Force's 1990s investigative efforts. Only one brief quote was included in the *Fact versus Fiction* report, cherry picked from another source. Does Colonel Marcel's partial omission from the US Air Force's 1994-97 reporting mean that their statements about finding "all the facts and bring them to light" might not be completely true? Remember the sentence I quoted earlier in this book from the *Case Closed* report? "Our objective throughout this inquiry has been simple and consistent: to find all the facts and bring them to light."

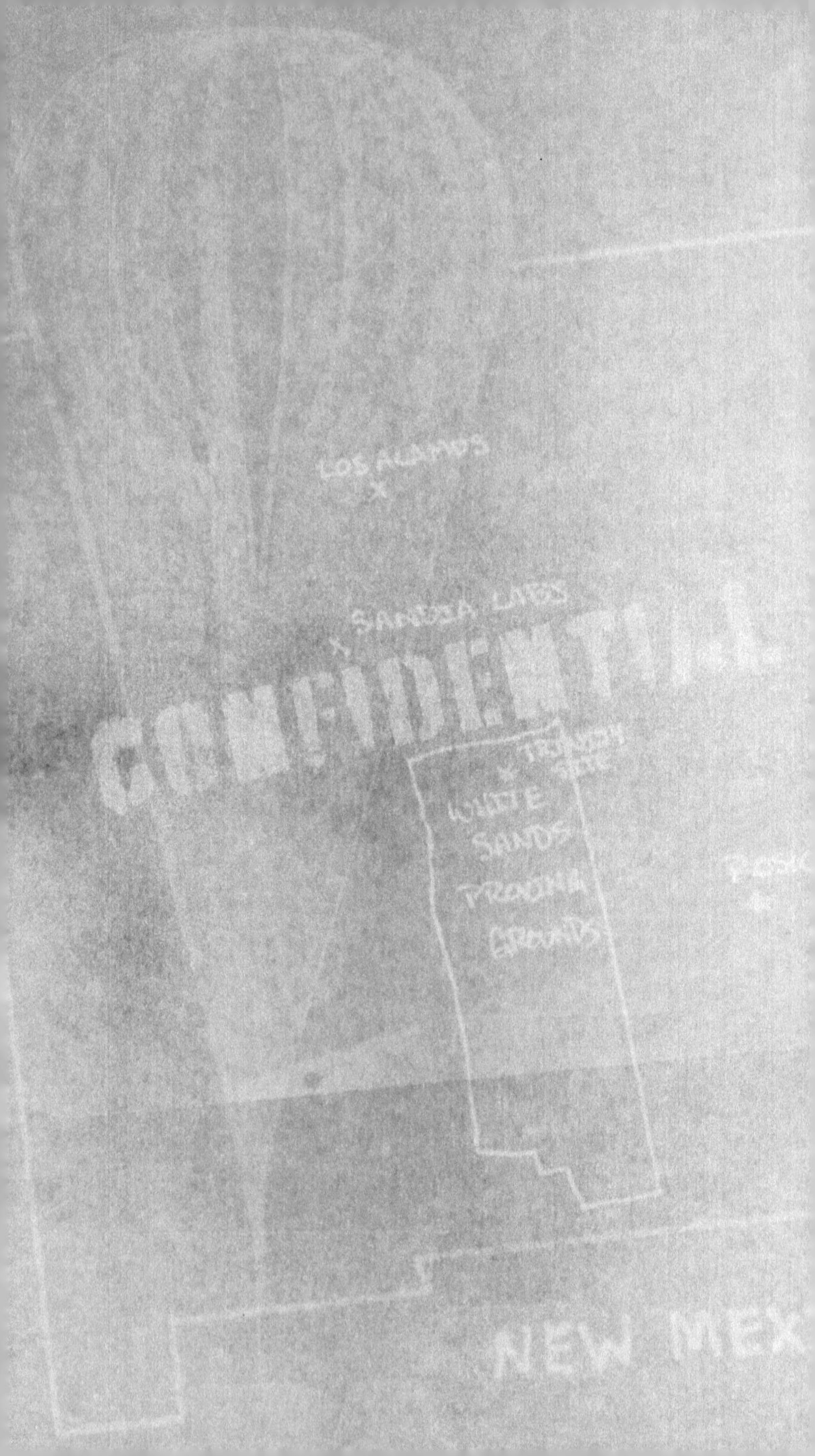
LOS ALAMOS
X
SANDIA LABS
X
CONFIDENTIAL
TRINITY SITE
X
WHITE SANDS PROVING GROUNDS
NEW MEX

CHAPTER 9
Summary and Conclusions

UFOs became a real, publicly discussed issue beginning in 1947. The reports collected and considered by the US Army Air Forces (the United States Air Force after September 1947) resulted in an assessment by Air Force Major General Twining that UFOs were real and constituted a threat to the US that the Air Force could not explain or counter at the time.

At the time of the Roswell Incident, the US Army Air Forces apparently did not have a public policy, strategy, or formal procedure to handle UFO Incidents. Colonel Blanchard issued a press release of his "find" in July 1947. The following day, General Ramey issued the "weather balloon" correction press release. A couple of months later, Major General Twining sent his classified memo documenting the US Air Force's need to address the UFO issue directly.

The need to protect information regarding the US Air Force's inability to overcome vulnerabilities made apparent by UFO/UAP was especially impactful to a branch of the military that was seeking independence and autonomy in 1947 and a nation that was assuming a global leadership role, including military power.

Science has moved from doubting the existence of extraterrestrial life to actively conducting research to discover it, spending billions of dollars each year to do so.

If extraterrestrial intelligence is interested in our planet, wouldn't the scientific and technical activities concentrated in New Mexico historically and in current times be a drawing card for them? Certainly in 1947 this would be especially true, with historic firsts in space, nuclear research, and radar.

During WWII and the years immediately afterward, the general population of the US was inclined to believe most of what the government and the military told them. And if they didn't, they knew their silence was part of the defense of the USA. In the twenty-first century this may not be the case.

After the Roswell Incident, the US Air Force began efforts to officially investigate and account for UFOs resulting in Projects Sign, Grudge, and Blue Book along with an effort at "scientific study" in 1968 with the Condon Report. None of these efforts found evidence of extraterrestrials. None of these reports included the Roswell Incident.

From a current view, one can easily surmise that the Air Force's efforts were not necessarily to investigate and seek truth, but to control, rebut, and discount any credible UFO information. It is a simple effort of logic to give motive to these activities: conduct counterintelligence activities to protect the information associated with vulnerability of the United States to extraterrestrial capabilities. I'll say it again: None of the reports included Roswell.

In 1994 and 1995, the US Air Force took charge of an "independent" Government Accounting Office investigation into Roswell begun at the request of US Representative Steven Schiff from New Mexico. The US Air Force's top counterintelligence officer led the investigation. This investigation and other USAF efforts resulted in the Air Force's series of official explanations of the Roswell Incident: first a flying saucer, next a weather balloon, later the 1947 top-secret Project Mogul, and even later the addition of anthropometric crash dummies from the 1950s.

Although the US Air Force no longer acknowledges the original flying saucer explanation, I included it in the above list of official explanations. General Blanchard reported it when he was on active duty, in an official capacity, and through proper procedure. He never recounted his story publicly. It stands as one of the official US Air Force explanations.

In Roswell, the annual July celebrations of the event that began in the 1970s have grown smaller and smaller while the UFO Museum and cottage industry souvenir shops languish.

Something happened in July 1947 in the New Mexico desert near the Roswell Army Air Field. The US Air Force has admitted they lied about it on at least one occasion (weather balloon), but mainly they have omitted the Incident from UFO/UAP efforts, even when investigating themselves in the series of 1994-1997 reports, concluding with the *Case Closed* report.

To discount the original stories of credible participants – Colonel Blanchard, Major Marcel, and Colonel Marcel – one must believe that none of these men could identify a weather balloon or a top-secret, data-gathering balloon that looked like a weather balloon.

The first and last official US Air Force personnel directly involved in the Roswell Incident were/are intelligence and counterintelligence personnel, all engaged in official USAF duties. Thanks to a seventy-five-year counterintelligence campaign, we are left with numerous thorny questions about what really happened in the New Mexico desert near Roswell in July 1947 – and it is challenging to sift through and align the "facts" that the US Air Force has publicly shared.

For example, because Project Mogul was a top-secret, compartmentalized program, Colonel Blanchard, Major Marcel, and Captain Cavitt probably did not know about it. But these competent, intelligent military leaders couldn't identify a balloon device when they saw it? To protect the classified information associated with Project Mogul, did the USAF immediately initiate a CI effort regarding the recovered material by declaring it was simply a weather balloon?

Did the US Air Force actually conduct an open, documented effort to determine what really happened at Roswell, resulting in publicly available reports surmising that top-secret Project Mogul was the basis of the Roswell Incident? In the investigation process that led to these reports, did the USAF honestly and diligently work to "find all the facts and bring them to light"?

Is there an ongoing counterintelligence campaign to protect the American people (and the world) from knowledge of tangible evidence of extraterrestrial intelligence, driven by the fact that this would represent an extraordinary vulnerability and threat to our airspace?

In the end, the evidence of an extraterrestrial encounter in July 1947 somewhere near Roswell, New Mexico, is more believable than the continued protestations, distractions, omissions, and counterintelligence campaign of the United States Air Force.

If this conclusion rings true, perhaps it is time to change the narrative about the Roswell Incident. I believe the US Air Force should lead the way, "Above All."

To continue the counterintelligence campaign with the current strategy and tactics will indicate that the Air Force is at odds with some 65 percent of Americans who believe there is intelligent life beyond Earth and that it is not a national security threat, according to a June 2021 Pew Research report.

It also confusingly indicates the Air Force is publicly in opposition to scientific searches for intelligent life beyond Earth – scientific searches that include the James Webb Space Telescope, Mars Exploration Rovers, and myriad unmanned spacecraft within and beyond our solar system. All of which are funded and generally supported by the American public.

No true American would fault the valid need to continue to protect information regarding vulnerabilities, capabilities, or other critical defense information. Just as the title of this book states, by applying an ongoing counterintelligence campaign the US Air Force is protecting us from the realities of Roswell. However, the time has come to share an accurate, historical account of what really landed in the desert near Roswell, New Mexico, in July 1947. The American people (and the world) deserve to know the *one* truthful story, so we can gain a clearer understanding of our universe, all military branches can team together to shore up vulnerabilities in our defense capabilities, and astronomers, astrophysicists, and other scientists can use the data in their search for extraterrestrial life.

So, for the seventy-fifth anniversary of the Roswell Incident, here's a suggestion for the United States Air Force: Show up in Roswell, New Mexico, on July 1, 2022. Send a General Officer, have a press conference at the Walker Air Force Base (formerly

Roswell Army Air Field), and announce that the US Air Force "can neither confirm nor deny the recovery of a flying disk in 1947." Better still, acknowledge the *one* truthful story of what really crashed in the New Mexico desert in July 1947. I will be glad to show you the town.

Or you can do the omission thing and simply continue to ignore Roswell.

APPENDIX 1

REFERENCES AND RESOURCES

PRIMARY US AIR FORCE DOCUMENTS REFERENCED, IN CHRONOLOGICAL ORDER

Letter, SEP 23, 1947, TO: Commanding General, Army Air Forces, Washington, D.C., ATTENTION: Brig. General George Schulgen AC/AS-2, SUBJECT: AMC Opinion Concerning "Flying Discs."

Unidentified Aerial Objects Project "SIGN," L. H. Truettner & A.B. Deyarmond, Classified: CONFIDENTIAL, (originally SECRET), 1949-02-01, AD0311102.

Project Grudge, Foreign Technology Div. Wright-Patterson AFB, OH, 1951-12-31, AD0128982.

Report of Air Force Research Regarding the "Roswell Incident," with Memorandum for Secretary of the Air Force, by Richard L. Weaver, Col, USAF, July 27, 1994.

The Roswell Report: Fact versus Fiction in the New Mexico Desert, Headquarters United States Air Force, , Colonel Richard L. Weaver and 1st Lieutenant James McAndrews 1995, ISBN 0-16-048023-X, US Government Printing Office.

USAF Fact Sheet 95-03, *Unidentified Flying Objects and Air Force Project Blue Book*, Current as of June 1995.

The Roswell Report: Case Closed, Headquarters United States Air Force, Captain James McAndrews, 1997, ISBN 0-16-049018-9, US Government Printing Office.

Additional Resources

"What Were the Mysterious 'Foo Fighters' Sighted by WWII Night Flyers?" *Air & Space Magazine*, August 2016.

"The Man Who Introduced the World to Flying Saucers," Megan Garber, *The Atlantic Magazine*, June 14, 2014.

Dragnet, "Just the Facts, Ma'am" quote from various online sources, e.g., imdb.com and snopes.com, February 19, 2018.

"RAAF Captures Flying Saucer on Ranch in Roswell Region," *Roswell Daily Record*, Roswell, New Mexico, July 8, 1947.

"Gen. Ramey Empties Roswell Saucer," *Roswell Daily Record*, Roswell, New Mexico, July 9, 1947.

Voice Recording, Interview of Sheridan Cavitt by Colonel Richard L. Weaver, National Archives and Records Administration, College Park, MD, Records Group 341-Stack Roswell-Row 3-Finder Aid Audio.

Voice Recording, General William H. Blanchard, "General W. H. Blanchard's Speech to Air War College," IRIS 905079, Reel 1, US Air Force History Agency, Maxwell Air Force Base, AL. (RSA Request 2015-00303, Richard Kestner, 12 Jun 2015)

Voice Recording, General William H. Blanchard, "Question & Answer to General W. H. Blanchard's Speech to Air War College", IRIS 905079, Reel 2, US Air Force History Agency, Maxwell Air Force Base, AL. (RSA Request 2015-00303, Richard Kestner, 12 Jun 2015)

COUNTER-INTELLIGENCE

Introduction to US Counterintelligence: CI 101 – A Primer, Mark L. Reagan, COL, USA (Ret), July 1, 2005.

Scientific Study of Unidentified Flying Objects, Volumes 1-3, Condon, Edward U., University of Colorado, 1969, available online at Defense Technical Information Center (DTIC).

The Roswell Legacy: The Untold Story of the First Military Officer at the 1947 Crash Site, Jesse Marcel, Jr. and Linda Marcel, Weiser, September 2008.

PDD/NSC-75, "US Counterintelligence Effectiveness for the 21st Century (U)," Dec 28, 2000, classified Confidential; unclassified fact sheet.

James M. Olson, "The Ten Commandments of Counterintelligence," CIA's *Studies in Intelligence*, Fall-Winter 2001, No. 11, pg. 54.

"An Assessment of the Evolution and Oversight of Defense Counterintelligence Activities," Michael J. Woods & William King; *Journal of National Security Law and Policy*, Vol. 3: No. 1, December 15, 2009.

Jung, C.G., *Flying Saucers – A Modern Myth of Things Seen in the Sky*, 1959, 1977 Edition, Routledge, Kegan Paul Edition.

Out There: The Government's Secret Quest for Extraterrestrials, Howard Blum, ISBN 0-671-66260-0, Simon and Schuster, 1990.

Sullivan, Walter, *We Are Not Alone: The Search for Intelligent Life On Other Worlds*, January 1966, Signet Books.

NASA, *SETI*, Government Publication No. SP-419, 1977, Updated August 10, 2004.

Billingham, John, Editor, *Life in the Universe*, NASA Conference Publication 2156.

Billingham, John, Project Cyclops. Moffett Field, CA: NASA CR 114445, 1971.

United States Air Force Official Website, Office of Special Investigations, March 10, 2022.

"Most Americans Believe in Intelligent Life Beyond Earth; Few see UFOs as a Major National Security Threat," Pew Research Center, June 30, 2021.

Joint Publication 1-02, Department of Defense Dictionary of Military and Associated Terms, November 8, 2010 (As Amended Through February 15, 2016).

Joint Publication 2-0, Joint Intelligence, October 22, 2013.

Joint Publication 2-01, Joint and National Intelligence Support to Military Operations, July 5, 2017.

Joint Publication 3-60, Joint Targeting, January 31, 2013.

Terms and Definitions of Interest for DoD Counterintelligence Professionals, Glossary (UNCLASSIFIED), Office of Counterintelligence (DXC), Defense CI & HUMINT Center, Defense Intelligence Agency, May 2, 2011.

Executive Order 12333 – United States Intelligence Activities, December 4, 1981, (As Amended Jul 30, 2008), National Archives, Office of the Federal Register (OFR).

Intelligence Community Standard Number 700-1: Glossary of Security Terms, Definitions, and Acronyms, Effective Date Remains: April 4, 2008, Director of National Intelligence.

"Preparing for Extraterrestrial Contact," Mark Neal, *Risk Management*, Vol. 16, No.2 (May 2014), pp. 63-87, published by Palgrave Macmillan Journals.

Backstory: Roswell: Exclusive Untold Disclosures About the 1994 Air Force Roswell Report Told By The Man Who Led The Inquiry, Richard L. Weaver, September 2, 2020, ISBN-13 979-8680329962.

Wilderness of Mirrors: Intrigue, Deception, and the Secrets That Destroyed Two of the Cold War's Most Important Agents, David C. Martin, July 1, 2003, ISBN-13-1585748242.

APPENDIX 2

GENERAL GLOSSARY

Words are important. Words like conspiracy and coverup are not used in this book. They are not the words accurately or appropriately describing the Air Force counterintelligence (CI) campaign to protect the information about the July 1947 Roswell Incident. Words like deception, distraction, inadvertent disclosure, need to know, cover story, disinformation, and omission are the language of the CI community and are used here extensively. Here are some of the important words, meanings, acronyms, and phrases used in this book.

United States Army Air Forces (USAAF) and United States Air Force (USAF) – In September 1947 the USAAF became an independent service, the USAF.

Roswell Army Air Field (RAAF) and Walker Air Force Base, Roswell, New Mexico – The same place. The name changed in January 1948.

Unidentified Flying Object (UFO) and Unidentified Aerial Phenomena (UAP) – Either one is not necessarily a flying saucer or extraterrestrial craft.

Intelligence – The product resulting from the collection, processing, integration, evaluation, analysis, and interpretation of available information concerning foreign nations, hostile or potentially hostile forces or elements, or areas of actual or potential operations. The term is also applied to the activity which results in the product and to the organizations engaged in such activity. (Joint Publication 1-02 and Joint Publication 2-0, Joint Intelligence)

Counterintelligence (CI) – Information gathered and activities conducted to identify, deceive, exploit, disrupt, or protect against espionage, other intelligence activities, sabotage, or assassinations conducted for or on behalf of foreign powers, organizations or persons, or their agents, or international terrorist organizations or activities. (Executive Order 12333, as amended July 30, 2008, and Joint Publication 2-01.2, CI & HUMINT in Joint Operations, 11 Mar 2011)

Threat – The intention and capability of an adversary to undertake actions that would be detrimental to the interest of the US. (Intelligence Community Standard 700-1, 4 Apr 2008)

Vulnerability – 1) The susceptibility of a nation or military force to any action by any means through which its war potential or combat effectiveness may be reduced or its will to fight diminished; 2) The characteristics of a system that cause it to suffer a definite degradation (incapability to perform the designated mission) as a result of having been subjected to a certain level of effects in an unnatural (man-made) hostile environment; and 3) In information operations, a weakness in information system security design, procedures, implementation, or internal controls that could be exploited to gain unauthorized access to information or an information system. (Joint Publication 1-02 & Joint Publication 3-60)

Countermeasures – That form of military science that, by the employment of devices and/or techniques, has as its objective the impairment of the operational effectiveness of enemy activity. (Joint Publication 1-02)

Classification – The determination that official information requires, in the interests of national security, a specific degree of protection against unauthorized disclosure, coupled with a designation signifying that such a determination has been made. (Joint Publication 1-02)

Classified Information – Official information that has been determined to require, in the interests of national security, protection against unauthorized disclosure and which has been so designated. (Joint Publication 1-02)

Confidential – National security information or material that requires protection and the unauthorized disclosure of which could reasonably be expected to cause damage to national security. Lowest Level of Classified Information. (Joint Publication 1-02).

Secret – National security information or material that requires a substantial degree of protection and the unauthorized disclosure of which could reasonably be expected to cause serious damage to national security. Middle Level of Classified Information. (Joint Publication 1-02).

Top Secret – National security information or material that requires the highest degree of protection and the unauthorized disclosure of which could reasonably be expected to cause exceptionally grave damage to national security. Examples of "exceptionally grave damage" include the revelation of sensitive intelligence operations and the disclosure of scientific or technological developments vital to national security. Highest Level of Classified Information. (Joint Publication 1-02).

Compartmentation – The principle of controlling access to sensitive information so that it is available only to those individuals or organizational components with an official "need-to-know" and only to the extent required for the performance of assigned responsibilities. (National HUMINT Glossary)

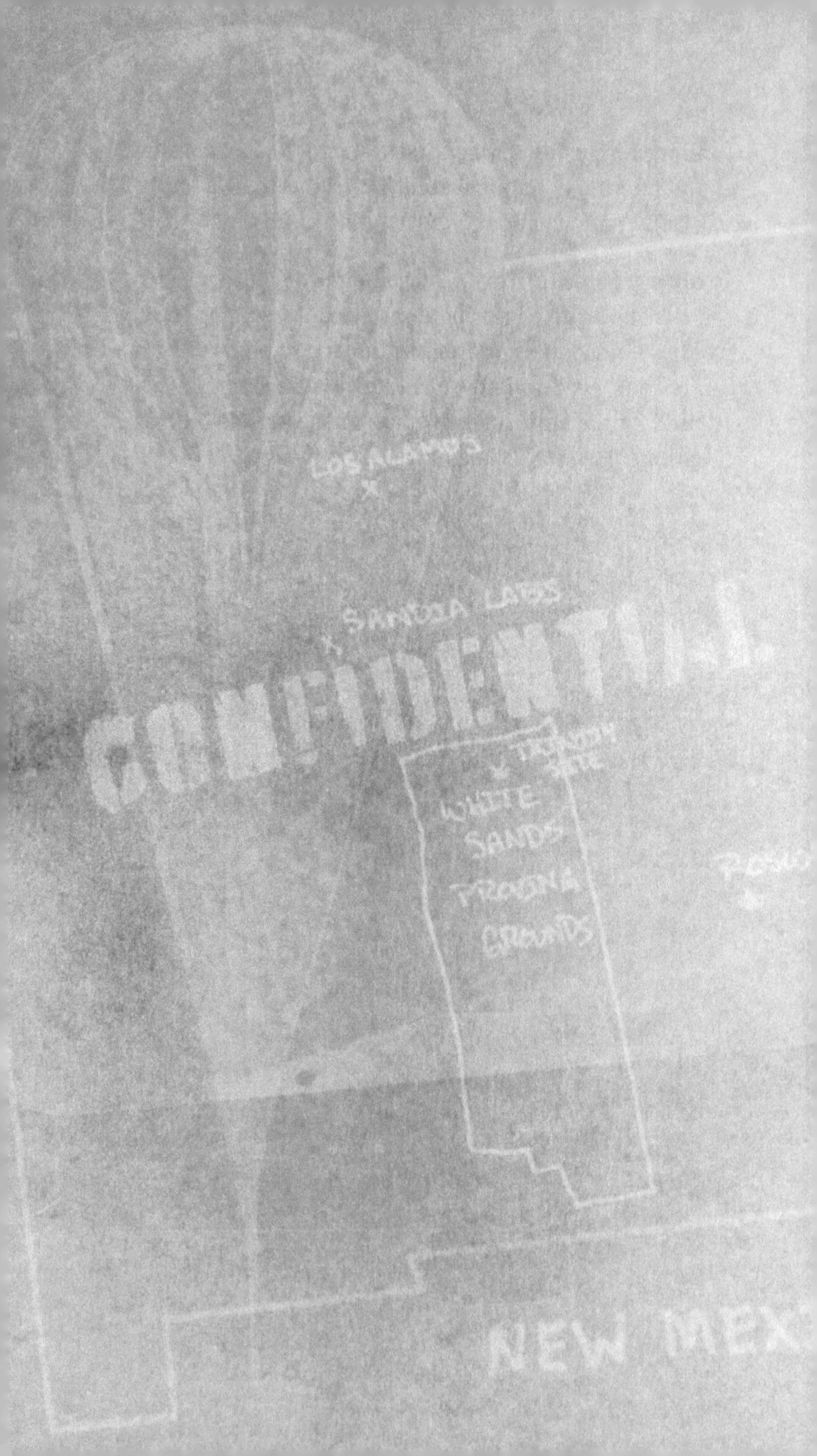

LOS ALAMOS
X
SANDIA LABS
X
CONFIDENTIAL
TRINITY SITE
WHITE
SANDS
PROVING
GROUNDS
ROSW
NEW MEX

ABOUT THE AUTHOR
RICHARD KESTNER

BORN IN 1948 at the Fort Belvoir, Virginia, US Army Hospital, Richard Kestner spent most of his life in, around, and somehow connected with the US military. As an "army brat" he was primarily educated by one of the greatest school systems on the planet – the Department of Defense Schools System, currently the Department of Defense Educational Activity.

Formative years were spent in North Carolina, specifically at Fort Bragg and Fayetteville. Graduation from Fayetteville Senior High School in 1966, a bachelor's degree from the University of North Carolina at Chapel Hill, and a master's degree from Florida State University completes his formal secondary and graduate-level education. In 1969 Richard "won" the only lottery of his life, #5 in the US Selective Service Draft Lottery. This prompted his enlistment in the US Navy Reserve in the "2x6" program: two years active duty and six years total commitment. The deferral of the active-duty obligation for a year allowed him to complete his undergraduate degree in Chapel Hill, North Carolina in 1970.

The US Navy recognized that a good number of highly educated individuals were enlisted in the same program, and they tapped them for active-duty assignment at the newly established US Navy Drug Rehabilitation Center (NDRC) at Naval Air Station, Jacksonville, Florida in 1971. The unit treated US Navy and Marine Corps personnel with addiction problems, mostly service members returning from duty in Vietnam. When Richard's active-duty obligation was fulfilled, he moved to Tallahassee, Florida, to continue in the graduate program begun at NDRC, graduating in 1975 from Florida State University.

First hired as the Deputy Director, Crossroads Drug Treatment Program in Fayetteville, North Carolina, Richard became Director after six months, increasing funding for prevention and education efforts when he renewed the annual Federal-State-Local grants.

In Spring 1976, he began his Civil Service career with the Department of the Army as an Education Center Counselor in Vilseck, West Germany, at the Seventh Army Combined Arms Training Center. After a couple of years and a promotion, he was personally recruited to head up an Army Mental Health Center/Community Drug and Alcohol Abuse Center in Amberg, West Germany, where he was part of the 3rd Squadron, 2nd Armored Cavalry Regiment, proudly living up to its motto, "Toujours Pret" (Always Ready).

When it came time to return to the United States in 1981, Richard chose New Mexico and was assigned to the US Army Training and Doctrine Command Systems Analysis Center (TRASANA), White Sands Missile Range (WSMR), New Mexico. Living in military housing at WSMR he was elected "mayor" of the community, working closely with and for WSMR Commander Major General Alan Nord in addition to his professional work. He received a promotion to the Army Materiel Test & Evaluation Directorate and broadened his work experience on many systems including many projects that were classified.

In 1985 he accepted another promotion to Director, Human Factors Laboratory, US Army Tropical Test Center, Panama. He took his family to Panama, served two years, and returned to WSMR when the political situation in Panama became tense in 1987.

From 1987 to 2005, he was a project engineer, conducting test and evaluation activities for a wide variety of military systems for all the services. Almost all the work was classified at some level.

For some projects, Richard was the only person at WSMR cleared for the project. He was competitively selected for his final civil service promotion to Team Leader of the Human Factors Engineering Team.

In the final years of his WSMR tenure, Richard was recruited to a newly established WSMR Marketing group, having been identified as a Project Engineer who worked mainly on projects he personally recruited to WSMR.

In 2004-2005, he teamed with the Office for Space Commercialization of New Mexico to submit a competitive proposal to the XPrize to host an annual XPrize Cup event. New Mexico triumphed over teams from California, Oklahoma, and the Kennedy Space Center in Florida in the competition.

Governor Bill Richardson asked Richard to assume the leadership role for the New Mexico Spaceport efforts in early 2005, and he retired from US government service. After jump-starting the Spaceport licensing process and recruiting Sir Richard Branson's Virgin Galactic space company to New Mexico, he retired again – unsuccessfully.

In 2007 Richard accepted a position with New Mexico Military Institute in Roswell as Institutional Research Officer. He was specifically hired to solve accreditation difficulties that NMMI was experiencing. Within a year, the Institute was back on track. Richard enjoyed being in Roswell and began work on his book about the Roswell Incident, staying until 2014 when he finally became successful at retirement. Richard published *Counterintelligence: Protecting Us from the Realities of Roswell* in 2022 – the seventy-fifth anniversary of the 1947 Roswell Incident.

Richard and Durema, his high school sweetheart spouse of more than fifty years, currently reside in High Rolls, New Mexico, where their home has a view of the White Sands, the Tularosa Basin, and the San Andres Mountains. Their two adult sons both live in Washington State with their families.

www.ingramcontent.com/pod-product-compliance
Lightning Source LLC
Chambersburg PA
CBHW031336060726
47590CB00007B/2494